RATS

1. The heads of the water vole (top left), the black rat (top right) and the brown rat (below).

RATS

MARTIN HART

Translated by Arnold Pomerans

Allison & Busby
London · New York

First published in this translation 1982
by Allison & Busby Ltd,
6a Noel Street, London W1V 3RB, England
and distributed in the USA by
Schocken Books Inc,
200 Madison Avenue, New York, N.Y. 10016

British Library Cataloguing in Publication Data
Hart, Martin
　Rats.
　1. Rats
　I. Title
　599'.3233　　QL737.R666
　ISBN 0-85031-297-3

All the photographs were taken by Hanneke van den
Muyzenberg except for Nos. 17, 19, 33, 44, 45, 46, 48, 49 and
50 which were taken by N. Gerrits; Nos. 5, 6, 7, 9, 10, 11 and
28 which were kindly supplied by the Dept. of Plant Pathology,
Vermin Control Section, Wageningen; Nos. 2, 3, and 4 which
were taken by J. Simons; and No. 8 which was kindly supplied
by the Print Room of Leyden University.

Set in 11/13 Garamond by Malvern Typesetting Services
Printed and bound in Great Britain by
Biddles Ltd, Guildford and King's Lynn

The rat of course I rate first. He lives in your house without helping you to buy it or build it or repair it or keep the taxes paid; he eats what you eat without helping you raise it or buy it or even haul it into the house; you cannot get rid of him; were he not a cannibal, he would long since have inherited the earth.

William Faulkner, *The Reivers*

2. The ratsbane seller. Engraving by Le Blond after a drawing by Abraham Bosse (seventeenth century).

CONTENTS

1.
ANIMAL OR MONSTER?

MY RATS

Long before I first heard the squeal of a rat, I had noticed the V-shaped ripples in the ditches along the smallholdings owned by my father and grandfather. They would say, "There goes a rat," when they saw the ripples, but stare as I might I hardly ever made out more than a large V or triangle in the water. Sometimes I thought I glimpsed a dark shape, a pair of small eyes or a set of whiskers, but the ditches were too narrow and the ripples disappeared too fast for me to be sure. In any case, I had no idea of what a rat looked like; I had never so much as seen a picture of one. And even had I spotted one in the ditches, I should have had a completely wrong impression because more likely than not those animals were water voles, not rats.

I never saw it. I was walking home from school under a grey November sky—I shall always remember the vast grey canopy stretching far above the shrieks of the rat. Sewerage men were working under the Jokweg in Maassluis; they had placed a small red-and-white fence round an open manhole and were busy below. Then, suddenly I heard a scream from one of the sewers, and a man rose out of the ground, a latter-day troglodyte for all I could tell, who, flourishing some iron object or other, was pursuing a shape that disappeared with improbable speed between the bay windows of two nearby houses. Within seconds, scores of people had gathered and I could hear the sound of iron on stone, together with so ghastly a scream that I started to shiver all over. A second sewerage worker came into sight carrying something on a shovel, but I could not see what it was because there were so many people in the way. From then on, whenever I heard the word "rat", the sound of a high-pitched scream would resound in my ears. I have heard this scream often enough since then and each time it freezes my blood.

Once I had heard the squeals, I did not find it so difficult to make sense of my father's oft-repeated story about his late grandmother. One day, about noon, she repaired to the chicken coop, spotted a rat in a corner and tried to catch it. The rat was cornered and in its plight, jumped up at my great-grandmother and bit her wrist. She then caught the animal by the neck, killed it with a pitchfork, and showed the two small wounds on her wrist to Henk, my great-grandfather, with a chuckle. During the afternoon a red weal came up on her arm and shortly before midnight she was dead.

And so, until the first time I was allowed to dissect a dead rat as a second-year biology student, the closest I had come to seeing one was a trail of ripples in a ditch and a crowd in the street, and all I had heard tell was a sad story that confirmed all my prejudices about an animal I had never seen. But the small white creature on the dissecting table had nothing in common with the dreadful monster from the sewers, nor had the live rats I saw two years later in the pharmacology lab. One spring evening, I bicycled to the laboratory with Helias Udo de Haes, who was to initiate me in the mysteries of etho-pharmacology. It was dusk and we passed under tall trees in which blackbirds were singing. He opened the door, we walked along several corridors, he opened another door, I saw the rats. They immediately stood on their hind legs and pressed against the bars, their little snouts protruding. It seemed to be a sort of greeting. The rats were extremely active in the dusk, much more active than they ever were during the day, de Haes told me, and that was why it was better to work at night. He prepared several large observation cages. While we were working I was startled by a loud noise followed by a splash of water. The bottom of the rat cages was covered with wire mesh through which the droppings fell on to iron trays which, as I later saw, were regularly tilted, so that a jet of water could wash away all the excrement.

We took the observation cages up into the loft, where hundreds of rats had been penned up. They raced round in their cages, having first greeted us, then tumbled over one another, fought, and coupled—males with males and females with females, for the sexes had been separated. There were no sounds other than a great deal of sneezing. It was an orgy of motion, and I have never tired of watching it since.

Helias Udo de Haes fetched the experimental animals. He placed a male on my palm. I stroked the animal. It was a queer sensation holding a warm rat in my hand—its fur was soft and I could feel its heart beat. On many evenings that followed Udo de Haes taught me to distinguish by name the various behaviour patterns of rats. After his departure, I continued to work with rats and, in particular, to study the effect of various drugs on their behaviour. The aim of our experiments was to discover more refined methods of identifying these effects than was possible with normal pharmacological procedures. We also hoped to draw conclusions about the action of these drugs on human behaviour. This is, indeed, permissible with several drugs—thus amphetamines have the same effects on rats as they have on human beings. Rats become agitated and antisocial, and so do people.

From Leyden I went on to work in Rijswijk. Although I would have liked to continue the same experiments—with other drugs—I was told to find out why the mothers in the rat pens rejected their young to such an extent that a shortage of experimental animals was threatened. And so I disappeared into the pens. I observed rats giving birth and investigated the behaviour of the mothers after parturition, and gradually it began to dawn on me why so many mothers were behaving in unmotherly ways. To begin with, males and females were being kept together in the pens and, because of the presence of the males, the suckling females were brought back into oestrus (heat). Now, a sexually receptive rat will reject her young if she is not fertilized. However, in Rijswijk the young were usually rejected immediately after birth, even before the females had a chance to come back into oestrus. Moreover, it appeared that many more young were rejected in the winter than in the summer—due doubtless to temperature and humidity differences. As soon as the central-heating system was switched on in the autumn, the number of rejections increased—the air was obviously too dry despite the presence of humidifiers, which probably had the opposite effect of what was intended. They released small droplets of cold water which acted as condensation nuclei for the water vapour in the air. The water

vapour condensed on these droplets, they became heavier and
fell down, with the result that the air became drier, not
moister. Moreover the humidifiers encouraged the spread of
bacteria. Only with steam could the air be properly humidified,
and on my advice the system was changed accordingly. As the
humidity rose, the number of rejections dropped to the summer
level. But this level, too, was abnormally high, so that another
factor had to be involved. I studied the immaculately kept
records. During the past thirty years the average number of
young per litter seemed to have increased gradually from six to
thirteen. How had this come about? In nature, large litters are a
disadvantage: as the number of young increases, each individual receives less milk and less attention from the mother. As a
result the young are enfeebled (remaining handicapped for life)
and, while they are still young, run a larger chance of being
eaten by owls and other predators. In nature selection thus
favours small litters. But in the laboratory members of large
litters have as good a chance of survival as members of small
litters, and because individuals from large litters make a larger
numerical contribution to the breeding of subsequent gen-
erations, we might expect to find that, in the laboratory,
selection favours large litters. This explains why the number of
young per litter born in laboratories increases from generation
to generation. Thus, in the Wistar Institute the average number
of young per litter increased from three to ten (Helen King). In
Rijswijk, I noticed that this trend had one very interesting consequence. With mothers who gave birth to, say, thirteen young,
the delivery of one young took up to ten minutes. The first-
born was left lying in the cold for up to twelve times ten
minutes while the rest of the litter followed. The young animal
grew stiff in the cold, just like its brothers and sisters who were
neglected for eleven or ten times ten minutes. When the whole
litter had been produced, the mother mistook the stiff animals
for corpses and consumed them. Now, while she was eating
them, the last-born were also left lying in the cold. They, too,
grew stiff and they, too, were consumed. As a result, selection
began to work against large litters. However, this was not the
only reason for the high number of rejections in the summer.
All the animals suffered from a serious pulmonary infection.

They sneezed the whole day long. After investigation by the bacteriological department and after the publication of my results, it was decided to destroy the entire supply of rats—a completely unforeseen consequence of my studies.

And so we stocked up with new rats. I learned a great many new facts about these animals and about experimental techniques. I studied the social behaviour of rats, their learning ability, the transmission of acquired behaviour patterns after brain transplants and their reaction to environmental changes. From all these studies it became clear to me that we knew very little about rats, that pharmacologists, animal psychologists and ethologists were all working in the dark. It was as if they were leafing through a book written in a language they did not understand, believing that they could piece together all the necessary information from the illustrations alone. Now the unknown language was the behaviour of rats, both in and out of the laboratory. Nothing about it was known. I accordingly decided to return to Leyden and to continue my behavioural studies, in the event with an animal that seems especially created for the ethologist, namely the stickleback. But in a small room I continued to keep brown rats which I firmly believe are much closer to their brethren in the wild than are white rats, if only because white rats have such poor eyesight. On many an evening I would again sit in front of my cages and look at the young at play, and at the aggressive and sexual behaviour of their elders, and, as I listened to the squeals of the fighting males, I dreamed of one day working with a team of colleagues and students and with the help of television and film cameras and good recording apparatus, making a thorough study of the general behaviour of these fascinating animals.

THE RAT: A MYTH

Visitors to our laboratory, seeing brown rats for the first time, would often say something like: "How small they are—I thought rats were much bigger." Other visitors might add: "Out our way they are *much* larger. Real whoppers." The speaker would then show with his hands just how gigantic these "whoppers" were—as large as a small pig, or even a small calf. I usually refused to let them get away with it: "Where did you see your rats?" I would ask. "Oh, well, I didn't actually see them myself but an uncle of mine lives near a canal and he always says . . ." or something to that effect. Occasionally, however, the visitor would maintain that he himself had seen rats in his own back garden, and from the size he indicated I usually concluded that it must have been the neighbour's cat or dog. My heaviest rat at the moment weighs 17 ounces and measures 9 inches from head to tail. The tail is another 7 inches long. The animal is not yet fully grown and will probably attain a weight of 20 ounces. Tables published by ecologists who have captured rats (pp. 49 and 51) rarely give weights of more than $17\frac{1}{2}$ ounces. Most rats weigh between 7 and 12 ounces and are no more than 7 inches long (plus 6 inches for the tail). Whence then the odd belief that rats are enormous animals? Is it that the largest rats have never been captured and is it precisely those rats that people usually see? They only see what cannot be caught—a most unlikely story. I am convinced that most people who see their first rat simply cannot believe that this friendly, small animal is the monster about which they have heard such terrible stories. They, like myself, have seen no more than ripples on the water and these ripples suggest a much larger creature. The legendary monster from the sewers simply has to be much larger than life. So much fear and horror cannot be invested in an animal that weighs less than 1 pound. The same strange myth also affects many other ideas about rats, culminating in the widespread misconception that a rat will fly straight at your throat. But the idea that the rat is a large, vicious and dangerous animal is not just a popular myth. Pest-control authorities reinforce it with their horrific posters—one, in particular, depicts a rat with a black mask and fearful claws, and a short television film about rats shows rats near a baby's cradle, ready to strike. This kind of propaganda is as spine-chilling as it is misleading. The animal depicted does not look at all like a rat, and the television film, in particular, is something I find utterly distasteful. Why not give sober information about rats? Why such vicious incitement? Is this really the only way to persuade people that rat control must be vigorously pursued?

But even famous writers and respected scientists convey a completely false picture of the rat. Let me give you a few examples. Chapter 10 of Konrad Lorenz's *On Aggression* is entirely devoted to rats. Lorenz has written magnificent books and articles but *On Aggression* must be considered something of a lapse. The chapter on rats is absurd and not only contains general untruths but, in particular, conveys a completely fallacious picture of the brown rat. What strikes one first of all is the tendentious choice of terms: "horrible brutes", "veritable murder specialists", "bloody tragedies", "victorious murderers", "one of the most horrible and repulsive things", "terrible fate", "sharp, shrill, satanic cry", "shrill war-cry", "group hate between rat-clans". These are not terms serious men should apply to animals. The theme of the chapter is that rat-clans are engaged in "constant warfare", "a form of aggression that we have not yet encountered: the collective aggression of one community against another". According to Lorenz this "group hate" does nothing whatsoever to further the preservation of the species. At the end of the book, he suggests, following a quotation from Goethe's *Faust* that the rat is some counterpart of man. Hellenius has paraphrased Lorenz's ideas in his *Plaatselijke godjes*, fortunately eschewing the more absurd terms, but nevertheless concluding that wars between rat-clans resemble wars waged by human beings.

Now, not a single scientific finding supports the existence of group hate between rat-clans. Lorenz took his main inspiration from Steiniger, whose descriptions are not nearly as colourful as his own. Moreover, anybody reading them carefully will see that the aggressive encounters between Steiniger's rats were exceedingly brief (2–3 seconds) and confined to individuals living in unnatural conditions. It is extremely difficult to tell precisely from Steiniger's more or less anecdotal accounts what he really observed, how many rats he used and how many experiments he performed. Working with free-living rats, such investigators as Telle and Ewer have, in any case, been unable to discover anything that can be interpreted as group hatred. Now it may well be that some such thing will be discovered in the future, for very few field studies have been conducted to date, but I, for one, do not think so. And even if such behaviour should be brought to light, our *present* state of knowledge about rats does not entitle us to write about these animals in the way that Lorenz has done. In any case, it seems most improbable that an animal as widely distributed over the face of the earth as the rat should be endowed with a quality that would lead to the internecine destruction of the entire species.

Lorenz also mentions that "knowledge of the danger of a certain fact is transmitted from generation to generation and the knowledge long outlives those individuals which first made the experience". Elsewhere he speaks of the "preservation and transmission of experience by means of tradition". This is a grotesque interpretation of isolated observations which should, at best, have persuaded him to make further observations, and never to utter such sweeping generalizations.

Novelists, too, have written incredible stories about rats. Thus in his *The Painted Bird*, Jerzy Kosinski describes what happened to a man who fell into a military bunker infested with rats: "His face and half of his arms were lost under the surface of the sea of rats, and wave after wave of rats was scrambling over his belly and legs. The man completely disappeared, and the sea of rats churned even more violently. The moving rumps of the rats became stained with brownish red blood. The animals now fought for access to the body—panting, twitching their tails, their teeth gleaming under their half-open snouts, their eyes reflecting the daylight as if they were the beads of a rosary."

But there is worse to come. "Suddenly the shifting sea of rats parted and slowly, unhurrying, with the stroke of a swimmer, a bony hand with bony spreadeagled fingers rose, followed by the man's entire arm." Further on, the author speaks of a "bluish-white skeleton", "partly defleshed and partly covered with shreds of reddish skin".

This passage is a modern variant of a tenth-century legend, according to which Bishop Hatto was locked up in the Mäuseturm in Bingen and eaten down to the bone by hungry mice. Kosinski has the story told by a child, which makes it even more horrible. The reader is spared no gory detail. This is the tenor of the whole book. I could only read it with a sardonic smile. Long horror stories cease to frighten when they consist of a whole string of gruesome descriptions and especially when the descriptions are so obviously false as are Kosinski's. He describes a bunker with starving rats that cannot escape. How did they get in there in the first place? The description, ("a sea of rats") suggests a large number of animals, as does the speed with which the man is stripped to the bone. Now, a large number of hungry rats consume one another, starting with the weakest, the next weakest, and so on until only one rat remains. Moreover, even starving rats are highly suspicious of food with which they are not familiar. They will never kill a human being. There is mention of gleaming teeth under half-open snouts. The boy looks down on the gnawing rats. How can he possibly see their teeth from that angle? Their eyes reflect the daylight. That, too, is a misrepresentation of what would have happened in the circumstances.

In his *1984*, George Orwell also uses fear of rats to conjure up nameless horrors. A cage with starving rats is about to be fitted over the face of a man, and the executioner delivers himself of the following address: "The rat . . . although a rodent, is carnivorous. You are aware of that. You will have heard of the things that happen in the poor quarters of this town. In some streets a woman dare not leave her baby alone in the house, even for five minutes. The rats are certain to attack it. Within quite a short time they will strip it to the bones. They also attack sick and dying people. They show astonishing intelligence in knowing when a human being is helpless." Here too we have the picture of rats that gnaw human beings down to the bone. Their ability to tell when a human being is helpless, incidentally, does not indicate an astonishing intelligence but only a lack of reaction on the part of the victim.

Orwell then goes on to tell how the cage is brought ever closer to the face of the victim. Here, too, there is the implicit assumption that hungry animals, which, moreover, are kept in cages (Orwell clearly knew something about rats), will attack a human being directly. If the hungry animals have been well trained, this is of course perfectly possible. But then Orwell goes on to write two highly equivocal sentences: "One of them was leaping up and down, the other, an old scaly grandfather of the sewers, stood up, with his pink hands against the bars, and fiercely sniffed the air. Winston could see the whiskers and the yellow teeth.'' A hungry rat does not jump up and down. It is impossible to tell by looking at a rat whether or not it comes from the sewers. A rat keeps itself perfectly clean. The word ''scaly'' is a misnomer, and it is far from easy to see the small yellow teeth. Even if they are seen, moreover, they are far from frightening. Orwell describes aggressively motivated rats. If the reader, incidentally, should ever be threatened in this way— and 1984 is not so far off any more—then he has three excellent ways of parrying the threat. He can blow at the rats, something they (and many other animals) abhor, he can scream or wail, preferably at a very high pitch, and he can bite back—enough to put any rat off.

The chief mistake in these (and similar) stories, for instance H. P. Lovecraft's ''The rats in the wall'', is the idea that rats chew corpses down to the bone. In Lovecraft's story the bones of the victim are found. Now, I have never read anything in any publication, not even in the books of rat-catchers written in past centuries, or in a host of newspaper articles about the behaviour of rats, to justify this strange belief. Those who do not defend themselves may lose their noses, a few fingers and toes, but will never be eaten down to the bone by rats. Moreover when rats do consume corpses (of conspecifics or other animals) they eat the bones as well. All that is left behind of other rats is the tip of the tail.

Rats are also used by many writers for atmospheric effect. A full account of this aspect of literature would need a whole book, and make monotonous reading at that: the rat is rarely presented as anything but a dismal and abhorrent pest. A case in point is Bordewijk's *Rood paleis*. He writes: ''Some sewer rats, half-hidden, with long, hairless tails, scratching away lethargically . . .'' (but sewer rats, i.e. brown rats, do not have long tails). And: ''There was a stench of stagnant mud. It was teeming with rats.'' He even mentions the discovery of a rat king: ''There was a whole band of rats, sewer rats with intertwined tails, all stuck together. These rats, so horribly welded together by a mysterious disease, constituted a king.'' In fact, no rat king has ever been discovered among sewer rats.

In his *La Peste*, Camus draws an impressive picture of dying rats which is correct in every detail. But then Camus writes about an animal, not about a monster, and without even so much as hinting at horrors he achieves more with a few sentences than Kosinski with a whole book. In Edgar Allan Poe's *The Pit and the Pendulum*, rats (with red eyes, albinos perhaps?) gnaw through a prisoner's straps. When they try to carry on gnawing, the prisoner drives them away. There is not a single improbable detail in the story (apart from the red eyes) and it is characteristic of Poe's genius that, while making use of rats to conjure up an atmosphere of fear and evil, he allows the animals to bring about a favourable turn in the fate of the prisoner.

Usually one has to guess whether writers or poets (for instance Swift, Trakl, Benn, Davies and others) are referring to black or to brown rats. But there is one literary work, a children's book, that draws a most favourable portrait of two rats and clearly distinguishes the brown from the black. The two wayfarers (one of whom, incidentally, loves to linger at home) in Kenneth Grahame's most beautiful *The Wind in the Willows* are called the Water Rat (brown rat) and the Sea Rat (ship's rat or black rat). There can be no objection to his portrait of the brown rat— the calm, hardworking animal Grahame describes would be welcome as a pet by anyone. The black rat, too, is magnificently described.

THE RAT AS A PET

During my stay in Rijswijk I came across one aspect of the rat with which I had not been familiar: the rat as a pet. Later I was to discover that Brehm, Barrett-Hamilton and Hinton had written eloquently on this very topic. In Rijswijk, laboratory assistants would often take superannuated rats home with them. One assistant even kept the offspring of a wild father and mother. It was a glorious brown animal with bright eyes, and grunted when it was crossed. A rat is normally much friendlier than a mouse, a hamster or a guinea-pig, animals that are often kept as pets. It likes to be picked up and stroked. Middle-aged women, in particular, will often overcome their initial fear of rats, to become quite besotted with these animals. An old lady I knew was inseparable from her rat. The animal accompanied her everywhere she went, although, when she first saw it, she gave a piercing scream. A tame rat is, of course, much more easily transported than a cat or a dog—it can be carried about in a lady's handbag.

Children, too, are usually very fond of rats. Not weighed down with all sorts of prejudices, they generally need no more than ten minutes before they pick up a rat and stroke it, as I have seen my nephews and nieces do time and again. Children like to push rats along in a doll's pram, covered with a little blanket. There is, however, one fairly strong objection to keeping rats as family pets: they are highly susceptible to pulmonary infections which can be communicated to children. But this seems to happen very, very exceptionally.

Because a rat eats everything that does not bite back or is not as hard as steel, the feeding of rats poses no problems, although, since rats are anything but vegetarians, they must be separated from other domestic pets such as tortoises and tropical birds. But then few people would leave a cat in an aviary. For the rest, a rat can easily be fed on table scraps and stale bread. It should never be offered sugar. Hamster food, which is sold in pet shops, is suitable provided it does not contain too many sunflower seeds, which are too rich in fat. One can give rats rare treats with bananas, chocolate, peppermints or pieces of soap; even alcohol (especially beer) is gladly accepted though, for normal purposes, it is enough to affix a feeding bottle filled with water to the rat's cage (the animal must have its own cage into which it can withdraw, and the door must of course be left open).

Female rats are more or less house-trained; they choose a special place for their needs and rarely go elsewhere. Male rats, too, can be house-trained but not nearly so easily because they are used to marking their runs with urine and faeces.

The rat is a social animal, even in its dealings with man. I cannot illustrate this better than by telling you the story of Kobus, the pet rat of one of my class-mates. Kobus lived in a wooden box, open only at the top. He kept gnawing at the sides, but not to make a round hole through which he might escape. Instead he made slits at more or less equal intervals, thus creating a kind of cage. He stopped gnawing as soon as he could look out.

There are many people who keep rats as snake food. On a Sunday afternoon the whole family gathers round the cage of their pet boa or python and watches while a live rat is thrown to, and devoured by, the hungry snake. No one to whom I have ever told this has shown any sign of indignation. But when I tell them that hamsters are used for the same purpose they usually protest. I am glad to say that I have heard of cases where the rat bit large chunks out of the skin of the snake.

The step from the rat as a chosen pet to the rat as an uninvited though welcome visitor is smaller than most people think. I know cases of elderly people (and others) who used (and continue) to feed wild rats by hand in their gardens every day. This may sometimes be the only daily encounter such lonely old people still have. I shall probably end my days in the same way. People like this are of course in a small minority. Far more often, man will wage war on rats with great persistence (see Chapter 8), much as he also breeds them with great persistence as experimental animals.

RATS AS EXPERIMENTAL ANIMALS

Those few who keep rats as domestic pets do not look upon them as monsters, but almost everyone else considers rats to be just that. One wonders why it is that an animal that fills so many people with fear and loathing and that has become the symbol of dread disease and evil should have become one of the most popular experimental animals of all. (Mice may be more popular still, but unfortunately we lack exact figures.) Is it an accident? According to Skinner, the investigator needs an organism that is readily obtainable and that can be kept cheaply. How many animals satisfy this demand? Only the brown rat (or the mouse)? No, many thousands of species. Why then the rat?

It is difficult to establish when rats were first used as experimental animals. Textbooks mention the year 1856, when

Philipeaux made a study of the adrenal gland. But Hooke and Boyle performed experiments with rats as early as 1660–70. Their book contains a fine illustration of a pump for removing the air from a glass cylinder. The cylinder contains a black rat standing against the wall of the vessel. The legend reads: "The behaviour of the rat while the air is slowly pumped out of the cylinder."

At that time, and perhaps even earlier, therefore, rats were already being used as experimental animals. But the heyday of rat experimentation came at the end of the last century. From 1877 to 1885 large numbers of rats were used by Crampe in his experimental studies of propagation. In 1894 H. H. Donaldson began to use them in Chicago; in 1906 he moved them to the Wistar Institute in Philadelphia, whence they have spread to laboratories all over the world. They are called wistars to this day.

Why rats of all things? Lockard believes it is because rats were used in a favourite sport of the English at the time. They would throw a number of rats into a pit, and then take bets on which of the rats would be the last to be killed by a fox-terrier (an exceptionally good ratter). Lockard believes that while breeding rats for this sport, people realized how easy it was to keep them, and that this may have suggested the idea of using them in laboratories. And once the tradition was started it was never broken again.

I cannot believe that this is the only explanation. Rats could not possibly have become laboratory favourites as the result of such chance developments. Anyone looking at the picture of the dying black rat in the cylinder cannot but feel that he is watching a ghastly spectacle. This description is true of many other experiments conducted with rats. But then rats are fair game; indeed, to most people, they are nothing but vermin. This, I believe, is the real reason rats have achieved such popularity as experimental animals. Those who kill vermin commit no crime. It also explains why people are not indignant about the use of rats as snake fodder while they become incensed at the use of hamsters for the same purpose. It is perfectly all right to have rats slaughtered by a terrier. Rats are vermin that can safely be done to death. Mice are treated in a similar way although, on the whole, they are not subjected to experiments nearly as horrible as are often performed with rats. Those men in white coats who pick up a rat by the tail and smash its head against a slab must have suppressed the feeling that they are dealing with a living creature.

Be that as it may, the rat has been foremost among animals used in experiments since 1900. The number of publications devoted to it is legion. In his famous textbook, Munn mentions three thousand publications in 1950; a more recent estimate gives more than three thousand articles every year. To keep up with the subject one would have to read ten publications a day. No reader will, I hope, expect me to have read all these papers—to have done so I should have had to read three publications a day since the moment I was born.

The experimental study of rats has two poles: pharmacological, medical and biochemical research on the one hand and ethological research on the other. Animal psychologists occupy a position somewhere between these two poles. Medical writers such as Munn look upon the rat as a "valuable instrument" or an "exceptionally useful tool". For these people the rat is a small bag of enzymes weighing some 7 ounces to which something can be added or from which something can be removed (or both) to produce certain results. If it were possible they would prefer to work with a test-tube (*in vitro* as it is called) full of enzymes. But in living beings (*in vivo*, as it is called), the enzymes are kept at steadier temperatures and better conditions in general so they are thrown back on the rat for better or for worse. The rat is still treated, however, as if it were a test-tube. The investigators freeze or heat it, they increase or reduce the pressure under which it is kept, they decapitate it (for which purpose they have a special guillotine) and, if necessary, they drown it. Some readers may feel so outraged by these disclosures that they will rush out to join the nearest anti-vivisectionist group. But some of the studies are of vital importance. If they were to be stopped, a great many important medical investigations would cease at the same time. The result would be a rise in human mortality, an increase in psychological tensions (for the development of psychotropic drugs owes a great deal to experiments with rats) and a host of other untoward developments. In this connection I might mention that some British anti-vivisectionists were responsible for releasing laboratory rats that had been used for the study of infectious diseases. These animals carried exceptionally dangerous germs and it was only because the escapees were white (and moreover sick) rats which perished quickly outside their pens that no terrible disaster ensued.

It is an undeniable fact that doctors do need a bag full of enzymes. They might perhaps be gentler with their experimental animals, use fewer of them, keep them in better cages, and stop that part of their work that is both cruel and pointless. But how is one to tell the good from the bad? Only after a research project has been completed is it possible to say whether or not it was worthwhile. Sometimes it takes years before a piece of work can be judged—before and during the investigations, perhaps, its real importance was unknown. If research were confined to the study of socially useful applications, as so many people demand, scientific work would be severely hampered and eventually dry up.

Ethologists, for their part, are interested in the rat as a rodent with an interesting and complicated behaviour pattern about which they want to learn more. They are concerned with the animal itself. Unfortunately, their publications are few and far between (an example of an excellent approach is R. F. Ewer's work with free-living black rats in Africa). Skinner says in one of his books that very few people are interested in the rat as such. They study the behaviour of the rat in order to draw conclusions about the behaviour of human beings. That is indeed how many animal psychologists think. They do not investigate the learning of rats but learning in general, not the aggressive behaviour of rats but aggression in general, and so on. Skinner has gone further in this respect than anyone else (see Chapter 9) and although his studies and those of many other animal psychologists are extremely important, I nevertheless think that they lack real depth, for none has taken the trouble to look at refuse dumps and river banks at nightfall with binoculars in the hope of learning something about the actual behaviour of rats in their (semi-) natural environment.

THE RAT: ANIMAL OR MONSTER?

I remember staying with friends on a houseboat, leaning out of a window one summer evening and looking across the waters of an Amsterdam canal. Suddenly, there were ripples in the water and then I noticed a mother with six young, all swimming in procession, the mother first, the young each in the triangle of its predecessor.

"Rats," I said.

My hostess wanted to close the window.

"Don't do it," I pleaded.

She leant out by my side. Just then the mother reached a broad plank that was drifting in the water. She climbed up followed by her young.

"How loathsome!" my hostess exclaimed.

"Why are they loathsome?" I asked.

The mother was cleaning her coat and so were the young. In the dusk, the preening young animals looked both beautiful and extraordinarily touching.

"They're horrors. Foul and dangerous, if you ask me. They carry diseases and they attack babies."

"Did you ever see them doing any such thing? If you hadn't been told all that by someone else would you still think them loathsome?"

She peered at the mother with her young. "Just look at those disgusting tails," she said.

"Do you think squirrels are disgusting too?" I asked.

"No, but they are altogether different."

"No," I said. "They are rats with bushy tails."

"Even so, those filthy creatures give me the creeps," she said.

"They are washing themselves," I objected. "They are neither monsters nor horrors; they are fascinating like all animals. The only reason they are pests is that we present them with refuse tips, with sewers, kitchen waste and lumber-yards in which they can multiply to their hearts' content. Individual rats are far from noxious, the harm lies in their large numbers. I wish someone would write a book about their behaviour, their way of life and their relationship with man. Perhaps then people would come to realize that rats are ordinary animals and not the monsters they are made out to be. But so little is known about them, much too little. It's just the same with owls and bats . . .."

Because I had worked myself up and had raised my voice, the mother stopped cleaning herself. She looked across at me, then jumped off the plank into the water, followed by her young.

"Oh, they do give me the creeps," said my hostess. "Let's close the window now, if you don't mind."

2.
A FAMILY AND ITS PAST

Rats are mammals belonging to the ORDER of rodents, or *Rodentia*, like a good half of all mammalian species. A characteristic of rodents is their long, rounded body on short legs. The hind legs are usually longer than the front legs; the trunk merges into the head without a clearly marked neck; the upper and lower jaws contain a single pair of chisel-shaped incisors which, as they become worn down at the tip, continue to grow from the base. If a rodent is prevented from gnawing, for instance if it is fed exclusively on milk, the upper and lower incisors will grind one another down. If the upper incisors are removed then the lower may grow into the nostrils. Next to these incisors, which are rootless, there is a large gap, called the diastema. It is situated just where humans have their canine and first molar teeth. To the rear of the mouth are the molars; their number varies from species to species. Many rodents, including rats, have three molars.

One possible way of subdividing rodents is the following:
1. *Sciuromorpha* (including squirrels, marmets, gophers and beavers);
2. *Myomorpha* (including hamsters, voles, rats and mice);
3. *Hystricomorpha* (including Old and New World porcupines); and
4. *Caviamorpha* (including cavies, chinchillas, coypus and water hogs).

These four SUBORDERS are divided further into FAMILIES, of which there are a total of 136. The families are divided into GENERA and every genus is subdivided into a number of SPECIES.

The suborder Myomorpha includes nine families, of which the *Muridae* contain the true mice and rats. One of the genera in this family is the genus *Rattus* which, according to the standard work, *The families and genera of living rodents*, consists of 544 different species (E. Walker even speaks of more than 570 species). Not surprisingly, therefore, this genus is the despair of taxonomists; there is still no general agreement about its precise division into species. What is certain is that the two species of rats prevalent in Britain, *Rattus rattus*, the black rat, and *Rattus norvegicus*, the brown rat, belong to this genus. Sometimes two other species are also included in lists of indigenous rats, namely *Rattus alexandrinus* and *Rattus frugivorus*, but these two are merely differently coloured forms of the black rat.

Before I come to black and brown rats, I should like to make a few comments about three rodents with which the brown rat is frequently confused, namely, the coypu, the musk-rat and the water vole.

THE COYPU (MYOCASTOR COYPUS)

The coypu is a large animal with a weight of some 20 pounds and a length of 16–25 inches, not including the tail. The tail is long and cylindrical and measures 12–18 inches. The animal has very large orange-coloured incisors and a ponderous head with small ears. The toes of the hind limbs are connected by webs (except for the fourth and fifth toes). The coypu is light brown in colour, and although the coat is fairly rough on top the soft undercoat has made the animal popular as a supplier of nutria fur. This explains why this shy South American animal began, in about 1930, to be bred in Western Europe. Quite a few escaped and gradually spread along the rivers. Coypus are good swimmers, they eat aquatic plants and dig simple galleries in river banks. They do not easily survive very severe winters, so that few were left after the frosts of 1963 on the Continent. In Britain, the coypu established itself in various places in the 1930s, but especially on the Norfolk Broads. During the late 1950s the animals began to spread widely over eastern England, damaging crops and river banks.

3. The ratcatcher of Rotterdam. Engraving by Cornelis Visscher
(seventeenth century).

4. The ratsbane seller. Drawing by Gavarni (nineteenth century).

THE MUSK-RAT (ONDATRA ZIBETHICA)

The musk-rat can grow to the size of a wild rabbit, and is thus much smaller than a coypu. Its weight ranges from 20 to 60 ounces, and its length from head to tail is 12–15 inches. The prominent tail is laterally flattened and 8–10 inches long. The ears are barely visible; the incisors are small and inconspicuous. The glossy fur is dark brown or chestnut on the top and lighter brown underneath, and is known commercially as musquash. The animal is well adapted to aquatic life. When it swims, a skinfold closes off the inner ear. The musk-rat is not averse to brackish or salt water and constructs very complicated systems of passages in dykes and dams. The advance of this animal into the Netherlands from Germany and Belgium has continued unabated despite the fact that special catchers have been employed by the authorities. In 1955 less than five hundred individuals were caught; in 1966 more than ten thousand individuals were caught; and since then the number of catches has increased every year. In Britain, musk-rats escaped from fur farms between the wars and have established themselves in the Severn Valley and elsewhere. Here, too, they were so destructive of river banks that they had to be expensively exterminated.

The original musk-rats in Europe are said to have been three pairs brought back by Prince Colloredo-Mansfeld from a hunting trip to Alaska. The prince, in his wisdom, is said to have released the animals, except for one male which died on the way, on his estate at Dobrisch, forty kilometres south-west of Prague.

The musk-rat has no natural enemies in Europe, except possibly the otter. Sevenster believes that the large sums spent on eradicating musk-rats would be much better invested in the protection of otters: a large otter population would effectively keep down the number of musk-rats.

The female musk-rat is edible; the meat is quite tasty but spoils very quickly. The male is inedible because of its musk glands.

THE WATER VOLE (ARVICOLA TERRESTRIS)

The water vole (see Fig. 1), also misleadingly called the water-rat, is closely related to the musk-rat but much smaller; it is no more than 5–7 inches long, and weighs $2\frac{1}{2}$–$6\frac{1}{2}$ ounces. It has been present in North-Western Europe for thousands of years; in the Netherlands, there may even be two species: the Limburg water vole does not live near water, and seems to be quite distinct from the water vole found in the west of the country. Water voles live along watercourses, and spread quickly, especially along regulated rivers. They construct deep galleries of their own or take over the nests of moles. They are very easily caught, all that needs to be done is to place a trap at the entrance of a nest.

Water voles have a marked preference for the roots of fruit trees, rose bushes, poplars and other trees, and also for tulip bulbs. They like vegetables and fruit, and can cause serious damage to gardens. They also carry many diseases, including a form of rodent plague that also attacks human beings. The animals may occur in very large concentrations in very small areas. Thus in Tomsk district, Siberia, more than four million water voles were caught within the space of just one month.

When people speak of water-rats, they may be referring either to brown rats or to water voles. The two animals are not at all alike, except that both are swimmers and are often spotted in the water.

SCHLEGEL ON BROWN AND BLACK RATS

The first detailed Dutch account of brown and black rats was Schlegel's *Natuurlijke historie van Nederland, gewervelde-dieren* (Natural history of the Netherlands, Vertebrates, 1860). It has some excellent descriptions and since it conveys a good idea of what was known about both species more than a hundred years ago, I feel justified in quoting from it at some length, the more so as it contains a number of mistaken views that continue to be held to this very day. Schlegel writes:

> The brown or rufous rat, which is often but unjustly called the water-rat, is the largest of all the species found in our country. It attains a length of $1\frac{1}{4}$ feet; the tail accounts for less than half this length, reaching, when stretched forward, as far as the ears, and is made up of some 200 transverse rings. The fur is a tawny-grey on top, growing lighter at the sides. The underparts, the feet and the inner side of the legs are white.

The brown rat first became known in this part of the world during the last century. Buffon reports that it was not seen in Paris before 1753. In Denmark it appeared during the last, and in Switzerland at the beginning of the present, century. And while it spread over Central and Northern Europe, it was also being carried by ships, in which it likes to nest, to all other parts of the world, forming colonies in many places from which, under favourable circumstances, it continued to migrate further. There is no certainty about the original home of the brown rat, on which subject we are forced to rely almost exclusively on Pallas, according to whom brown rats are widespread to the South-East of the Caspian Sea, where they build their nests in holes made and later deserted by porcupines. In the autumn of 1727, large hordes of rats reached the banks of the Volga from the Western desert, crossed the river, infested Astrakhan on the other bank, and devastated everything they could seize. All this happened a few days before an earthquake that was observed all along the Western bank of the Caspian Sea, and since, shortly afterwards, the Plague caused great depredations in the above-named city, the appearance of rats was considered by many of the more credulous a premonition of this disaster. Once again migrating from the West to the East, similar hordes of rats also appeared in Gur'yev at the mouth of the river Ural or Yaik, shortly before 1770, at the time that Pallas was travelling in these parts. All such reports would seem to suggest that the brown rat, together with many other species of mice, left its home at a time when an extraordinary multiplication in its numbers took place, for regions in which it could find better sustenance; that similar hordes repeatedly migrated from the European side of the Caspian Sea, along its Northern shores, to the Asiatic side during the first half of the last century; that other rats probably migrating in a Westerly direction, much as happened during the repeated migrations of the nations, spread all over Europe. Rats do not seem to have migrated to the North of the Caspian Sea, since neither Pallas nor any other traveller encountered them in Siberia; however, at the time Pallas visited the area, they were already very numerous in St Petersburg. In Southern Europe, to the North of the Alps and as far as Geneva, there could be found another brown rat with a longer tail, namely *Mus alexandrinus* or *tectorum* i.e. *Rattus alexandrinus* [a sub-species of the black rat, M.H.] which seems to have come from North Africa, where it was first observed in Egypt.

In Europe, the brown rat infested human habitations, especially in the larger cities, in which sewers afforded it safe hiding places and where all sorts of refuse and an abundance of food allowed it to multiply in great profusion. The black rat, which the newcomer encountered on its arrival in such places, had to make way before the invader and has as good as disappeared from many areas. The brown rat likes the proximity of water, swimming and diving most proficiently, which is why it is generally known as the water-rat. It is as voracious as it is bold, and will even attack human beings when it is cornered. It eats anything it can seize, be it of animal or plant origin. It plunders nests, devouring eggs and young birds; it kills young chicks, drags young ducks under water to drown them, strips turkeys, and, if necessary, does not spare its own kind. The female has two litters a year, indeed sometimes as many as three, of four to eight young each, which are born blind.

Many people, and especially women, consider rats to be repulsive creatures, which does not, however, prevent them from flaunting the skins of these animals, which in Paris are caught in incredible numbers for this very purpose, in the form of kid gloves.

The black rat is a little smaller and less sturdy than the brown rat but its tail is longer and consists of some 250 rings; the ears are larger and take up half of the head as against only one third in the brown rat. The coat is dark brown and this colour, although it grows slightly lighter and shades off into grey, also extends across the nether parts of the animal which are white in the brown rat. In this species, just as in the house mouse, one often observes albinos which can be tamed and interbred for the purpose of providing pets or for public display.

In our parts the black rat is never met in its original wild state. We may therefore take it that it, just like the species we named earlier, must have reached us from abroad and that this event probably occurred in the Middle Ages, since the writings of the ancients make no mention whatsoever of this animal. The first naturalist to give a recognizable description and to mention its occurrence in Germany was Albert the Great, who lived in the twelfth century. Its original home is unknown, but it is thought likely that the animal reached us from Asia. During the past century, it was very numerous in cities and villages and caused so much damage in some places that Church bans were pronounced over it and days of penance were proclaimed. To the extent, however, that the brown rat began to spread across Europe and to wage war to the death on the black rat, the latter was almost completely eradicated during the first quarter of the present century. Since that time it has either become very scarce or has completely disappeared. On the other hand, once having taken to ships like the brown rat, it was conveyed to other parts of the world, where it spread in various regions.

In this respect of its habits, manners, propagation and choice of food, this species shows the greatest degree of affinity with the brown rat. However, it does not choose to live near water and will not swim if it has the choice.

Here and there, although very rarely, a pile of ten or more dessicated rats has been found in a confined space, all joined together by their intertwined tails, after apparently huddling together and perishing for lack of food, the exits of their holes having been closed in one way or another. To such groups of rats the people of Germany have given the name of *Rattenkönig* (rat king) and this phenomenon is observed not only among black but also among brown rats.

Although Schlegel's account of brown and black rats compares very favourably with the fables of so many later authors, it, too, contains a number of errors. To begin with, Schlegel was misinformed about rat kings (see Chapter 4); the similarities between brown and black rats are far fewer than he suggests and it is a moot point whether the brown rat has in fact ousted the black.

The reason why neither the brown nor the black rat was mentioned in the "writings of the ancients" was not that neither was known but rather that there was a lack of linguistic distinction between the rat and the mouse. In the Bible, for instance, there is just one word for a certain rodent (a rat or a mouse?) which in English translation is usually given as mouse (Leviticus 11:29; 1 Samuel 6:4, 5, 11 and 18; Isaiah 66:17). In classical literature, too, there are references to a rodent which the Greek described as μῦς . Herodotus refers to the field-mouse as μῦς αρουραῖος .In Greek there was also a word for an animal (υραξ) which was later called *sorex* by the Romans and which clearly referred to the water shrew. Usually it is impossible to tell from the text whether the writer refers to a rat, a mouse or some other rodent, just as with modern writers who, when writing about rats, usually do not make it clear whether they are referring to the black or to the brown species. Herodotus reports that rodents gnawed through shield-straps and bowstrings when Sanherib advanced on Egypt. These rodents must surely have been rats. Indeed rats must have been so widespread in countries round the Mediterranean Sea even then that we must simply accept that when they spoke of rodents, Greek and Latin writers were frequently referring to rats.

In Epinal's glossary (A.D. 700) the rat is not mentioned. Archbishop Aelfric mentions "raet" in the *Vocabulary* he compiled in about A.D. 1000, but this word may have been derived from the Provençal, in which language "rata" means the house mouse. According to Hamilton and Hinton the first writer to distinguish explicitly between mice and rats was Giraldus Cambrensis, who lived from 1147 to 1223.

Nor is it at all certain that the brown rat did not reach Europe before 1700. In his *Historia animalium* (1553) Konrad Gesner depicts a rodent that looks suspiciously like the brown rat. If I am right, then the brown rat must have reached Europe before the middle of the sixteenth century.

THE BLACK OR SHIP RAT (RATTUS RATTUS)

It is widely believed that the black rat entered Europe on the ships of the returning Crusaders. Thus, in all sorts of late mediaeval works, we can read reports about an animal that could hardly have been anything other than a black rat. In the fourteenth century the species was already so widespread and caused so much damage that poison was used to try and eradicate it. Rat-catchers and pedlars of ratsbane were popular figures from the thirteenth to the nineteenth centuries, giving rise to a host of fables and legends, among them the famous story of the Pied Piper of Hamelin. Pedlars of ratsbane were also the subject of many engravings (see Figs. 2, 3 and 4) especially in the seventeenth century. Illustrations and paintings by Rembrandt, van Ostade, Jan Steen, Pieter de Bloot and Cornelis Visscher, amongst many others, show men carrying very long sticks attached to a basket or box full of rats. The animals are always black rats. Sometimes there is also a dog, often a fox-terrier, and an apprentice of sorts. The rat-catcher on Visscher's famous engraving has a well-filled money-bag. In the seventeenth century, rat-catchers were not always mere beggars, although Rembrandt's prints certainly convey an impression of poverty. In other periods, however, rat-catchers were usually penurious or crippled (Figs. 2 and 4).

probably
anthrax
47-48)

The black rat carries the rat-flea, responsible for the spread of bubonic plague (see p. 138), the Black Death' which had Europe in its grip from the fourteenth to well into the eighteenth century. (Brown rat also carries plague)

The black rat is now increasingly rare (though it can maintain itself in unusually large numbers in certain localities, e.g. in London). This is true of the northern part of the northern hemisphere. Elsewhere, black rats are still very common, or as Corbet put it: "The black rat is the dominant villain in the world."

The black rat is a very beautiful animal, particularly when it is genuinely black, for there are all sorts of coloured varieties. Its pointed snout, long tail and larger ears help to distinguish it from the brown rat.

THE BROWN OR COMMON RAT (RATTUS NORVEGICUS)

The English name "brown rat" is better than the Latin, for the brown rat is not a native of Norway where, in fact, it was first observed towards the end of the eighteenth century. The German name *Wanderratte* (roving rat) refers to its frequent migrations, but such migrations also occur among other rodents. The Friesian name of *ierdrat* (earth rat) characterizes the animal as a hole-dweller. The name sewer rat is misleading because many brown rats do not live in sewers. The name water-rat should be avoided because this name causes confusion with the water vole. In England the species is known by the following names: brown rat, grey rat, common rat, wharf rat, water-rat, barn rat and Hanoverian rat. The last name arose because the brown rat was said to have entered England in the ship that brought George I from Germany in 1714. According to another old legend, the brown rat first came over from the continent with William of Orange in 1688 (Charles Waterton, *Essays on Natural History*).

It is believed that the brown rat originated in Asia. In which part of that continent? India and Persia are mentioned, and the species is said to have been in Mongolia since time immemorial.

In China, various species of short-tailed rats have been found that more or less resemble brown rats. We are told, moreover, that similar animals also occur to the west of Lake Baikal, but I am inclined to conclude from all the many conflicting reports that the original, wild form of the brown rat can no longer be found anywhere in the world. The brown rat as we know it today is a highly domesticated animal just like the dog, the cat or the horse. It is an animal moulded by us, and transformed by us. We have, in fact, bred a creature with an incredible capacity for adaptation. It is probable that the genuine brown rat has long since died out.

The brown rat can reach a maximum length of 12 inches (the figure incidentally, does not tell us very much about the animal because we may come across it either at full stretch or crouching, and the difference can amount to more than 6 inches). The tail of the brown rat, according to the taxonomists, has a maximum length of 9 inches. I myself have never seen a tail (or measured one) more than 7 inches in length. The number of rings in the tail is rarely more than 180.

Barrett-Hamilton and Hinton say that brown rats rarely weigh more than 17 ounces. They mention a reliable report of a rat weighing 53 ounces and they also cite two unreliable reports of rats weighing 29 and 31.5 ounces respectively. From the tables published by Chitty, Telle, Davies and others it would appear that rats very rarely weigh more then $17\frac{1}{2}$ ounces.

There are also very divergent data about the life-expectancy of rats. In the laboratory, white rats rarely live for longer than two years. From various investigations by ecologists it would appear that rats in the wild rarely live for longer than one year. The claim that rats can grow older than six years, as Richter suggests they do, is something I consider a fable. Richter himself gives no data to support this high figure.

Experts also disagree about the number of young produced by one pair during any one year. In the next chapter I shall discuss what few data there are about the fertility of rats; here I shall merely try to estimate the number of descendants of one pair under optimum conditions. On average, the number of young per litter is six; of these six, three are usually female. The gestation period is twenty-one days. Suckling lasts for another twenty-one days. A female can, however, be fertilized while she is still suckling her young. She can even be fertilized on the day of her confinement. For convenience, I shall put the period between two confinements at forty days. If, then, a female gives birth to six young on 1 January, she is able to produce another litter forty days later. The female offspring of the first litter of six are themselves capable of producing young within another hundred and twenty days. If I may take it that each litter consists of three females and if I count all the descendants of all the females in one year I arrive at 1,808 rats on 1 January of the following year, including the original pair. This is, of course, an improbably large number. First of all there are deaths; mothers sometimes reject their young; females often do not come back into oestrus for a long time. Nevertheless this figure should give the reader some idea of the army of rats that can arise from two individuals within only one year.

WAR TO THE DEATH

Schlegel wrote that the brown rat waged a "war to the death" on the black rat and Selma Lagerlöff in her *The Wonderful Adventures of Nils* describes how the black rats were conquered by the brown (she says "grey"). The grey rats, she writes,

> were descended from a couple of poor immigrants who landed in Malmö from a Libyan sloop about a hundred years ago. They were homeless, starved wretches who stuck close to the harbour, swam among the piles under the bridges, and ate refuse that was thrown into the water. They never ventured into the city, which was owned by the black rats.
>
> But gradually, as the grey rats increased in number, they grew bolder. At first they moved over to some deserted and condemned old houses which the black rats had abandoned. They hunted for their food in gutters and dirt heaps and made the most of all the rubbish that the black rats did not deign to worry about. They were hardy, contented, and fearless, and within a few years they had become so powerful that they started to drive the black rats out of Malmö. They took from them attics, cellars and store-rooms, and starved them out or bit them to death, for they were not at all afraid of fighting.

After this introduction, Selma Lagerlöff goes on to describe the fight between grey and black rats in a country estate, from which Nils Holgersson charmed all the grey rats away with his flute. I often think that it was from this great children's book that so many people got the idea that the brown rat has ousted the black. And perhaps it was this very fable that ran through Lorenz's head when he wrote his chapter on rats. Barrett-Hamilton and Hinton put it all much more cautiously: "The ousting of the black rat may have been in part due to a direct antipathy between the two species, and in part to the great voracity of the brown rat which perhaps tended to deprive the weaker species of provisions."

There are other opinions as well. A. M. Husson says: "One may postulate that the brown rat ousted the black rat, but this assumption completely ignores the fact that ousting presupposes a struggle for territory and food. But this was not the case. The brown rat has a preference for damp cellars and basements, for sewers, pigsties, refuse dumps, ditches and canals, while the black rat prefers a dry habitat." F. E. Loosjes, too, has argued (in an article published by *Rat en Muis* in March 1956) that the brown rat could not have ousted the black rat because their respective choices of food and biotope are quite different.

What are the facts? After 1700 the number of black rats gradually decreased and the number of brown rats quickly increased. Is there a causal nexus between these two developments? Or is there simply the same coincidence as exists between the increase in the use of electric razors (from which, incidentally, the clippings are inhaled) and the increase in lung cancer? Was there so marked a change in the natural conditions of brown and black rats after 1700 that one species could increase while the other species waned without the one attacking the other? The increase in the number of brown rats coincides; in any case, with the spread of sewers and increasing urbanization.

Another argument frequently produced in support of the ousting hypothesis is that, when brown rats are locked up in cages with black rats, they will invariably kill the black. Thus Robert Smith, the rat-catcher of Princess Amalia, already explained, in *The Universal Directory for taking Alive and Destroying Rats, etc.* (1768), that he put Norway rats, which he had caught in a cellar, in a great cage together with black rats, which he had caught in the attic of the same house, in order to show these rats to his employer, the owner of the house. But as soon as the brown and black rats met, the brown rats attacked the black, killed them, and devoured them. This is an interesting story. Smith caught both species in one and the same house. Until the moment he caught them, the brown rats had not yet ousted the black. It is quite possible, therefore, that both species had not yet met each other. Many other observers report the same fact: peaceful co-existence of brown and black rats in different parts of one and the same house or building.

Brown rats will also kill house mice and water voles if they are placed in the same cage. The killing of mice by rats is, in fact, the basis of many pharmacological experiments which, by the way, strike me as singularly absurd and pointless. Now, outside these cages, neither the house mouse nor the water vole has been eradicated or driven out by the brown rat even though, as far as choice of biotope and food are concerned, they differ no more from the brown rat than does the black.

The question of ousting is an extraordinarily interesting one that is far from being solved. Since the 1950s, the Turkish turtle-dove has been spreading in the Netherlands while the common turtle-dove is becoming increasingly rare. Has the one been ousted by the other? Everywhere in Western Europe and particularly in Britain, the grey squirrel is increasing in number while the red squirrel is decreasing. Has one ousted the other, or is it rather that an open place, a niche, in the language of ecologists, has been occupied by the grey squirrel, in which case the red squirrel must have begun to disappear before the grey squirrel increased? Again, such animals as the willow warbler and the chiff-chaff, the curlew and the whimbrel may occur in the same territory but not in precisely the same spot. This seems to resemble the situation of brown and black rats, which may indeed live within the same house but in different parts—one in the cellar and the other in the attic. Is the situation of the brown and black rat comparable to that of the Turkish and the common turtle-dove or to that of the curlew and the whimbrel? Before we can answer this question, we shall have to study both species far more thoroughly than has been done so far.

H. Dienske has shown with a series of beautiful experiments that field-mice will oust field-voles, but no such experiments have been conducted with brown and black rats. There is thus no firm evidence that the first species has ousted the second.

3.
INTRUDERS

On many an afternoon and evening in the early summer I used to lie in the long grass round refuse tips and the banks of stagnant pools watching the incessant toing and froing of brown rats. Provided I did not move, they paid hardly any attention to me. Sometimes they would even sniff at me. What these rats were doing differed little from what other animals do when man observes them in nature: they were looking for food. In the laboratory, rats have food in abundance so that they have a great deal of time for other activities, but in nature the search for food is the main daily task, which explains the relative scarcity of aggressive and sexual behaviour. Even the playing of young animals is something I only noticed on a gigantic rubbish dump in Rijkswijk, where there was an obvious profusion of food. It is quite possible that wild rats in their holes behave just like rats in the laboratory, continuously mating and fighting, but they cannot, of course, be observed in their holes. However it seems most unlikely that they would behave in these ways—after all, searching for food all night long is not the same thing as idling about in a cage before being confronted with another rat.

By saying this, I in no way wish to suggest that, since animals in nature behave differently, laboratory investigations are meaningless. If rats fight on a rubbish tip they fight in the same way as laboratory rats; it is just that they fight less frequently. If we want to study the structure of fighting behaviour, laboratory observations will do very nicely; indeed, we know that caged rats fight more frequently than free-living rats, i.e. that they provide more aggression in a short time. But from the greater quantity of aggression we are not entitled to conclude that rats are aggressive animals. In nature they are peaceable enough. Even the smooth-walled cellars measuring just under 700 square feet in which Steiniger kept his wild rats and which he called a *Freilandgehege* ("free enclosure") were in fact large cages, and

Steiniger's observations of a strict ranking order among non-related rats, of the severing of carotid arteries and of pair formation are therefore not at all indicative of what happens in nature.

Those who observe rats in nature are primarily interested in their relationship with the environment. This is the province of ecology. A host of ecological data has been collected by Barnett, Davies, Davis, Leslie, Steiniger, Veenables and others. A synthesis of their results is not a simple matter because their findings are often incompatible—a systematic consequence of the fact that rats behave differently under different conditions.

THE HOMES OF RATS

Black rats originally lived and nested in trees, and in warm countries they still build their nests in that way. In this part of the world they never burrow in the ground like brown rats. Hans Joachim Telle did, however, observe black rats living in burrows both in North Africa and in Syria. In Western Europe, black rats seek out dry places (for instance attics) but in Mediterranean countries they are also found near water. Brown rats can build their nests wherever the surroundings are not too dry and good access to the nest is assured. They like to be well covered and, in particular, to sit out under a kind of roof. Thus the verandas of ground-floor flats offer them ideal cover, the more so as puddles and low shrubs often go with this type of "desirable" habitation in new districts. But wood-piles, bales of straw and corn ricks also provide them with good nesting sites. When no such shelter is available, they will dig holes in the ground. In these burrows, a whole group of rats will live together, and I shall refer to such groups, mostly families, as packs. Telle has established that of the 1,401 packs he studied, 524 lived outside (37 per cent) and 877 in buildings (63 per cent). Outside, brown rats chose moist places without exception; inside they may also use dry spots (for instance the vicinity of a boiler).

Outside, rats construct complicated systems of passages down to a depth of 18 inches. Steiniger, who has made a special study of the subject, has divided burrows into three types: living areas, larders or eating areas, and hiding-holes. The difference is probably merely a phase in a constantly changing situation. The larders are usually located between the hiding-holes and the living areas; as soon as the distance between living area and larder becomes too great the larders are converted into living areas. The same thing can happen to the hiding-holes.

The living area has three to five, or, if the pack is very large, up to eight, entrances which join up after a run of $1\frac{1}{2}$–$4\frac{1}{2}$ feet. Close to their junction lies a chamber which nearly always gives on to a small blind alley. This blind alley is relatively deep (10–14 inches) and several yards long. It generally widens at the end. When digging up rats' burrows, Steiniger almost invariably found the animals in these passages with their heads towards the blind end. It appears that it is extremely difficult to pull rats out by their tails, which are usually amputated in the process. The inhabitants of a German North Sea island told Steiniger that the rats creep into these blind alleys during floods and close them up with their bodies (the passages narrow down so that, at a given point, a rat fits into them like a cork). As a result the air cannot escape, and the rat is said to stay alive as long as the oxygen lasts.

The larger the packs, the bigger and more complicated are the burrows. On the North Sea island of Norderney (with a total rat population during Steiniger's investigation of 5,000) Steiniger discovered passages more than fifty yards long. Often there are numerous living chambers and blind alleys and occasionally there are passages so narrow that they must have been dug out by young animals. The larders are very like the living areas. When they are dug out, rats or nesting material are rarely found in them; there is instead a great deal of food. Larders do not give on to blind passages. Hinton (1920) found that most larders are eventually converted into living areas. The hiding-holes are no more than short passages close to the sources of food. The rats disappear into them when large birds fly overhead, or human beings approach, and so on. These holes undergo a constant process of reconstruction; new entrances are built, old entrances are abandoned and allowed to collapse, all the entrances being changed in the course of two months. The internal system of passages and chambers, too, is constantly being altered.

Brown rats leave their homes in a characteristic manner. One animal comes out first, stops at the entrance of the burrow for about two minutes, generally standing on its hind legs, though not necessarily all the time, and tests the wind (see Fig. 12). If all is clear, it will leave the hole, followed fairly quickly by the other animals. Barnett reports that it is always the same rat (the pioneer rat) which is the first to appear. Telle observed that different rats would come out first on different days and that, in large packs using many exit holes, various rats would come out to test the wind simultaneously.

It is a striking fact that black rats will test the wind for much longer periods than brown, sometimes for up to thirty minutes (see Fig. 5). Among them, too, it is often one individual that sits or stands in the characteristic attitude, eyes half closed (Fig. 5), or, if squatting, head held low (Fig. 6). After having remained in this position for a long time and satisfying itself that everything is safe, this black rat is followed without hesitation by all the other black rats, none of whom stop to test the wind.

Whenever brown rats leave their burrows (or their nests in buildings), they invariably use the same runs. On this point most investigators are agreed. Black rats are much less tied to their runs than brown, but their home range (see below) is much smaller. Usually the runs of the brown rat are easily identified: whenever possible they provide cover from the side and from the top, for instance they lie along walls, fences, steep banks, and so on.

Telle has conducted interesting experiments which show how closely rats are tied to their own runs. He discovered two small packs (a small pack being a group of less than twenty rats) that shared the same territory. He destroyed all the members of one of the packs and put out six rats imported from a different area. The imported animals, which he marked, used the runs of the dead animals as well as the runs of the surviving pack, who defended their runs most vigorously. After several weeks the six marked animals were found to use only the dead pack's runs. Telle repeated the experiment three times in various places, always with the same result: two packs can share a territory without mutual aggression, provided that each animal confines itself to the runs of its own group, easily recognized since the runs are marked with the animals' droppings and urine. Runs and burrows together form the territory of a pack of brown rats, and that territory is defended against members of another pack. Brown rats, however, stray far from their holes in order to seek food, the area they cross in the process being called the "home range". The home range is thus larger than the territory, at least for brown rats. In the case of black rats in Northern Europe, home range and territory coincide.

How large is the territory of brown rats? Davies studied the rat population in Baltimore for many years (I use the word population when an author does not make it clear whether he refers to one or to several packs). He fed the rats a blue dye so that, from the distribution of the blue faeces, he could determine how far the rats had moved from their nesting places. Blue faeces were found within the limits of a circle with a diameter of 30 yards. This was a very small home range. Steiniger has reported a very much larger home range. At ebb-tide, he claims, rats will travel many miles over the mudflats in order to plunder fishermen's nets; he also observed animals that covered many miles in order to empty eel baskets. Telle, too, speaks of rats covering several miles.

In 181 of the 212 packs of black rats he investigated Telle found that home range coincided with territory. Among the remaining packs of black rats the source of food lay just outside the territory, a fact clearly connected with the black rat's choice of habitat. Thus if a pack of black rats inhabits the loft of a barn in which food is stored, then the loft constitutes the combined territory and home range of this pack.

DAILY RHYTHM

Rats are said to be nocturnal animals, but this is not quite true. "Rats", Telle asserts, "can usually be found outside their holes at such hours of the day as are particularly quiet." Thus rats can often be observed at the end of the lunch hour and in the early evening. But here too there are exceptions. "In many refuse dumps rats can be seen from sunrise until nightfall." According to Telle, rats are particularly active at sunrise. On a refuse dump in Rijkswijk I myself observed that rats invariably came out and proved very active during lunch breaks when no human beings were normally about. This is probably a very general phenomenon. Rats regulate their activities to fit the circumstances. In restaurants, for instance, they are only active after closing time. As soon as human beings stop work, the rats will come out. In offices and factories they remain active from closing time until 8 p.m.; then they rest until the next morning. As Neuhaus has stressed, however, their activities fluctuate with the seasons: in winter rats rise later and retire earlier than in the summer.

ANNUAL RHYTHM

In the autumn, brown rats are said to leave their burrows to enter buildings; in the spring they are said to re-emerge and to dig new burrows. Davies, Herold, Steiniger and others mention this annual migration, although Herold himself asserted as early as 1935 that rats preying on field-mice stay outside in the winter. Steiniger speaks of a spring and an autumn migration; his striking descriptions cannot, however, disguise the fact that he adduces few tenable arguments in support of his thesis. ''For example,'' he writes, ''one can often see rats in the country appearing at intervals of a few hours or half-days, alone or in small groups, as if they were following a scent. They follow precisely the same runs, and cross ditches in precisely the same way.'' How does Steiniger know that these are different rats? They might very well be the same rats in search of food, taking a different route out from the one they take home. And how often has Steiniger observed them doing so? How does he know that they take this route only in the spring and autumn, not in the summer or winter as well? As long as we have no further information we cannot possibly conclude as to an autumnal migration. The spring migration, Steiniger himself admits, cannot be observed directly, but must be deduced from the difference in the number of rats in the same building in winter and spring. That argument, too, is not very convincing. Telle, who has also examined this problem, went to Bülkau, a village in Germany, and recorded the migrations of brown rats as accurately as he knew how. He later repeated his observations in the village of Delmenhorst. This is what he found: in the autumn there is indeed a migration; groups of three, four or five rats were found in buildings that had previously been free of rats. However, outside packs also grow larger during the autumn as migrating animals join existing packs, so that groups of three to five rats were discovered in open places that had previously been rat-free. The migrating rats were probably young animals born in the spring that had meanwhile reached sexual maturity and had been driven out of the original territory by the older animals because there was not enough food for all. Similar developments have been observed in many other species of rodents, especially among young males. This accounts for the autumnal migrations (in the autumn there are many young, sexually mature animals). This type of migration is in no way a trek from the outside to the inside of buildings but must be attributed to population growth in the spring and the summer. Moreover, in the autumn food becomes scarcer so that the migration of young animals is encouraged even further. The young animals are chased out. In the spring, though, there is no perceptible migration from buildings to the outside; indeed, during the summer there are more packs inside buildings than during the autumn and winter, or so Telle claims. Outside, too, there are more packs in the spring and the summer than in the winter. This must be due to population increases in the summer.

Brown rats do not hibernate but during very severe winters they keep to their burrows. Experts are not agreed on whether or not rats stop producing young in the winter. In laboratories there are usually far fewer pregnancies in the winter, but if temperature and humidity are regulated so as to simulate summer conditions, the number of pregnancies nearly reaches the summer level. In the packs studied by Telle the percentage of pregnant females decreased after June. In April, May and June, 40 per cent of the females were pregnant; during subsequent months the percentage of pregnant females dropped by an average of 6 per cent per month, until in December, January and February less than 5 per cent of the females were pregnant. This was followed in March by a very marked increase in the number of pregnancies. Because the number of pregnancies is smaller in the summer than it is in the spring, there is thus no correlation with the temperature level. In sewers, where the temperature is low but where temperature fluctuations are very small (in fact the temperature is almost constant) no differences in the number of pregnancies in the summer and winter were discovered.

In many other species, the number of daylight hours is a very important factor in fertility fluctuations. In rats there are no indications that this factor causes an increase in the number of pregnancies. In the sewers it is always dark and food is present most of the time. Perhaps this is an important element. In laboratories and granaries, too, there is a good supply of food the whole year round. Even a small degree of underfeeding in the laboratory is reflected in a change in the sexual receptivity of females: instead of coming into oestrus every five days, they may only come into oestrus every five weeks or so. For the rest, rats will propagate their kind under very unfavourable circumstances—the crucial factor is the food supply. Both Steiniger and Gaffrey found pregnant females and nests of young in cold-storage plants in which the temperature was -8° to -10°C (18°-14°F).

POPULATION DYNAMICS

Population dynamics studies the factors governing the density of organisms in various places and at various times. It also deals with the average life-expectancy of a population, births and deaths, the ratio of males to females, population genetics (in so far as possible) and distribution patterns. If one wants to gain a better understanding of, say, such mysterious processes as the appearance of plagues of field-mice and the migration of lemmings, then such studies are of the utmost importance.

The first thing to establish is the size of the population under investigation. This is particularly difficult in the case of rats. Telle was one of the few to make accurate counts; most other investigators have contented themselves with estimates. Leslie and his collaborators, who caught and killed all the rats in the granaries they investigated, also produced accurate counts. They discovered 2,879 rats in 51 granaries. Each of the granaries was a population centre (that is, the centre of a pack), so that there was an average of 66.4 rats per pack, a size that commonly occurs among rat packs. We have just said that the number of births is dependent on the supply of food, the age of the females, and often also on the time of the year and the size of the population. Most investigators found an average of 6.2 young per litter. In the above-mentioned granaries an average of 9 young per litter was born, perhaps because of the abundant supply of food. The females in the granaries produced an average of 4.8 litters per year. These females were involved in propagation for less than two years; during their entire lives they produced an average total of eight litters.

Leslie divided the females in the granaries into five classes by weight. He would have liked to divide them into five age classes, but a field worker cannot possibly tell the age of captured animals with any degree of accuracy. However, since rats grow at a fairly constant rate, there is a very good correlation between age and weight. Table 1 shows that the average number of young per litter depends on the weight (and hence on the age) of the mother. The older a female the larger the litter. Females weighing more than $15\frac{1}{2}$ oz. have generally ceased to produce young; such females are, however, very rare. In the laboratory, too, it is very often found that each successive litter of a given female is larger than the preceding one.

Table 1. Weight of females and average size of litter (Leslie)

Weight	Number of females	Average size of litter
0–115g (0–4 oz)	no pregnant ♀♀	
115–195g (4–7 oz)	38	6.368
195–295g (7–10½ oz)	184	8.201
295–395g (10½–14 oz)	183	9.010
395–495g (14–17½ oz)	79	9.658
495g and above (17½ oz and above)	12	11.000

5. A black rat testing the wind. The eyes are half shut. The large ears show up particularly well in this photograph.
6. A black rat testing the wind with its head held low.

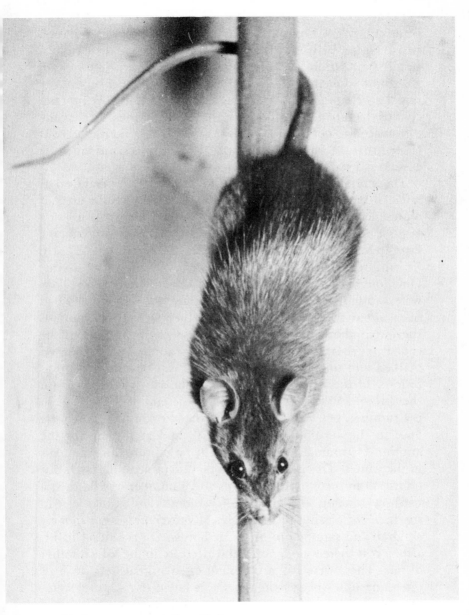

7. Black rats are outstanding climbers.

With ample food, a population will increase at a rate of 8–10 per cent per week. However, if there is not enough food to feed a constantly growing number of animals then the population will increase fairly slowly. David Davis removed rats from a stable population (122 from a population of 430 rats in Baltimore). One month later he found that 48.3 per cent of the females were pregnant. At the time of the capture only 14.4 per cent of the females had been pregnant. He established that the difference was not due to purely seasonal fluctuations, and concluded that the number of pregnancies increases quite suddenly as soon as a stable population is reduced.

The mortality of rats is difficult to determine. A great deal of research has been done in laboratories but the results are not representative of what happens in nature. I know of just one large-scale study of rat mortality in nature. Davis, working on a farm in Maryland, tried to establish how many rats out of 1,036 he captured, marked and released were still alive after a year. Only four of the 1,036 rats were recaptured after 46 weeks. They were two females and two males. Because the two females had been about twenty weeks old when they were first captured, they were about 66 weeks old at the second capture. In the course of these 46 weeks Davis also captured and counted marked and unmarked rats. His result can be summed up as follows: of the 1,036 rats captured at the start of the study (in the spring) 750 rats were still alive at the end of the spring. In the summer, 600 were still alive; in the late summer there were 540; in the autumn there were 270; and at the end of the autumn 150 animals were left. Finally he captured fifty animals in the winter. This is a fairly grim picture (at least for rats). But on this farm there lived four dogs and a number of wild cats; a barn-owl took up residence for a few months and skunks were in the habit of passing by regularly. Moreover, rats were shot by children and stamped to death by horses. On the other hand, after a year there was no noticeable decrease in the *total* number of rats. There were still a thousand or so animals left. It is an open question whether the rats Davis failed to recapture were, in fact, dead or left untrapped. In any case, Davis discovered many dead rats that had been marked, and it seems very probable that most of these rats lived for less than a year. The same thing probably happens elsewhere as well—unless the rats are spared by owls, rat-catchers and others who try to eradicate them.

There are many laboratory data but few field data on the sex ratio of rats. King found that the number of females newly born slightly exceeded the number of male, but Leslie's rats produced a slightly larger number of male descendants (51.87 per cent). However, because females live longer in nature than males, the ratio changes in favour of the females within just one generation.

Leslie and his collaborators were also the only ones to investigate the age structure of a large rat population. In fact they analysed the weight structure because, as we said earlier, the age of a captured rat cannot be determined with accuracy. The weights of the 542 rats they examined are given in Table 2. The table shows that the sample contained a large proportion of young animals.

THE SIZE OF PACKS

The difficulty of determining the size of a pack lies in the fact that the rats have to be counted, a far from simple process. Much has been written on the subject. Leslie and Davis, relying on certain regularities, have developed a mathematical model. They set down a very large number of traps and from the changing daily ratio of empty and full traps they computed how many rats live in a given territory. Now, because a certain number of rats are captured and removed every day, the chances

Table 2. **Number of rats in a given territory by weight and sex** (Leslie)

Weight	Males	Females
0–44g (0–1½ oz)	106	93
45–95g (1½–3½ oz)	21	32
95–145g (3½–5 oz)	39	38
145–195g (5–7 oz)	18	17
195–295g (7–10½ oz)	31	42
295–395g (10½–14 oz)	35	53
395–495g (14–17½ oz)	10	6
495g and above (17½ oz and above)	1	0

of catching rats in these traps diminishes daily—there are fewer and fewer rats to run into the traps. Conversely, the more rats there are the smaller the daily changes in the number of captures. From the speed of the changes in the relative number of empty and full traps it is thus possible to estimate the total rat population. With very large populations, say of more than a thousand rats, this is a good but laborious method: scores of traps have to be checked every day. Since, however, rat populations of more than a thousand animals are few and far between, different procedures have to be used. The most accurate method of all is undoubtedly to destroy all the rats and to count the corpses, but then there will be no animals left for further experiments. Nor will it do to capture rats, to mark them, to let them go and then to compute the population from the ratio of marked to unmarked animals caught on successive days, for this method only works if the chances of trapping marked and unmarked animals are equal. However, rats do not usually run into the same trap twice in succession. Another method—counting the number of rats that can be seen during the period that the animals are active—only works with small packs which are easily surveyed. The counting of the number of droppings, holes, rat-runs and the like may give the trained investigator a fair picture of the total number of rats present, but it, too, is not very accurate.

There is, however, one method that seems to have proved both relatively simple to use and fairly reliable. The rats are offered the kind of food they like and which they cannot scatter about: say, oats sprinkled with salad oil. The food is provided daily and abundantly. On the first day the rats will consume a certain amount of the food and on successive days more and more of the food will be consumed until, one day, saturation point will have been reached. That means that all the rats present, including the most cautious, are now taking the food. On average, rats consume 15 grammes of food per day or 10 per cent of their body weight, and the average weight, because a large number of young is always present, is 150 grammes. Chitty has used this method to count the number of rats in a population. Telle, too, has used this method. To check its efficacy, Telle used it to compute the number of individuals in thirty-eight packs, and then captured the members of the thirty-eight packs in spring traps. The number of rats caught and the number determined by the food method agreed very closely, at least with packs numbering less than eighty individuals.

Table 3 summarizes Telle's results for 1,401 packs of brown rats and 94 packs of black rats.

Table 3. Size of packs (Telle).

	Total number of packs	Size of packs (in classes of 20 animals)										
		20	40	60	80	100	120	140	160	180	200	220
Brown rats												
In buildings	877	144	203	153	90	86	32	66	26	1	9	67
Outside	524	86	112	75	60	56	39	37	28	5	6	20
Black rats	94	35	32	16	1	5	0	1	3	0	0	1
(only found in buildings)												

Packs should not be considered as super-families since stray animals may be absorbed into them (see below). Nevertheless some family connection is preserved even in larger packs. Mothers with young usually have their own living areas, which they retain. Since "fathers" play no role in family relationships, we speak of maternal families as they also occur in many other species of mammals.

INTRUDERS

When roving rats enter the territory of an established pack they are usually driven off or killed. At least twelve authors have described this process in some detail. Steiniger has drawn a most dismal picture, although the fights he observed rarely lasted for more than two or three seconds. Barnett has described it all more objectively but this conclusion is much the same: alien rats are kept out of the pack. Now all these reports deal with caged rats and the assumption is always made that things are the same in nature. However Hinton and British rat-catchers from previous centuries had already stressed the peaceable behaviour of free-living brown rats. Intruders are sniffed at, sometimes there is a brief skirmish and then the stranger is accepted (I myself have seen this happen twice). If the rover is a female she is immediately mounted, although sometimes other females will try to oust her. If she takes to her heels, she is never pursued.

In nature, packs are generally larger than those kept in cages or in "free enclosures". Is it possible, then, that the rigid defence of territories which undoubtedly occurs in the laboratory, is no more than a by-product of the loss of freedom by a fairly small group, just as the strange behaviour of apes on "monkey hills" in the zoo is a by-product of their aritifical environment. Laboratory rats have a rigid social hierarchy: dominant males are the first to have access to the females and to food, sometimes, as in Calhoun's experiments, to the exclusion of all other males. From Calhoun's experiments it appeared that non-dominant males could only feed when the dominant males were asleep.

As far as I can tell, Hans Joachim Telle was the first to make large-scale studies of wild, free-living rats. His name has been mentioned before, but now that I want to discuss the most striking results of his work I must say something about his investigations in general. From June 1952 to May 1964 he made observations with packs of black and brown rats in various parts of Germany. From time to time he would have trouble with farmers who, not unjustly, accused him of feeding rather than killing the rats on their farms. To pursue his studies of black rats, Telle travelled through Southern Europe and North Africa, the Middle East—all regions in which the black rat is prevalent.

Telle's results contradict almost all those obtained by laboratory investigations. Wild rats are less aggressive than rats in cages. Telle's picture of the brown rat is so convincing precisely because it is so much easier to imagine the animal as a successful rover and emigrant than as the monstrous brute it appears to be from the stories of Barnett and Steiniger (not to mention Lorenz).

Telle was able to show that the larger a pack the smaller is the likelihood of an intruder being killed or expelled. Table 4 summarizes Telle's results. There are three possibilities: the intruder is expelled; the intruder is accepted without further ado, or after a fight; or the intruder is killed.

A (±) after the weight of the experimental animal indicates that the animal was killed. Weights in italics indicate that the intruder was driven off by several animals. The figures in brackets in the last column indicate the number of animals that were only accepted after a fight.

Various interesting facts emerge from the table. In a pack consisting of less than forty animals there is a good chance that the intruder will be expelled; if he is not expelled he usually has to fight before he is accepted. Very small packs consisting of less than twenty animals invariable expel all intruders. Only young animals (weighing less than 200 grammes) are killed. Very exceptionally, several individuals will band together to drive out an intruder. In very large packs the chances of an intruder being expelled are very small and the intruders are usually accepted without a fight. Black rats behave in much the same way: large packs accept strangers.

In other words, the fact that laboratory workers never deal with large packs tends to distort their views of aggression. A pack of twenty animals is an exception in the laboratory; laboratory rats know one another individually, so that there is not the least need for a strange rat to have a strange scent—a stranger is recognized as such directly.

Among black rats Telle discovered a similar picture (see Table 5), though no animals at all were apparently killed. However, working with eleven packs that counted less than thirty individuals each, Telle discovered that as soon as one individual began to ward an intruder off two to four other individuals came to the assistance of the first. This happened in eight out of eleven occasions. Telle's findings agree with those of Mrs Ewer, who studied the behaviour of black rats in Africa.

When Telle released a strange brown rat in the territory of a small pack, it took some time before the stranger abandoned the trap Telle had opened. In one case the brown rat did not leave the open trap at all. More usually, however, it was driven out by a member of the resident pack that happened to pass by the trap. Once out of the trap, the strangers would move about

Table 4. The acceptance and expulsion of intruders by brown rats.

Size of pack	No. of intruders	No. of experiments	Weight (in g.)	Result expelled	not expelled
10	1	4	180,140,210, 120(±)	4	
10	2	1	280, 300	2	
10	2	1	170, 230	2	
10	4	1	170, 210, 170, 90	4	
10	1	2	280, 170(±)	2	
10	1	4	160, 170, 90(±), 170(±)	4	
15	1	4	120, 200, 210, 340	4	
15	3	1	100, 180, 80(±)	3	
20	1	4	110(±), 220, 240, 80	1	3(2)
20	2	1	130, 110(±)	2	
20	2	1	170, 130		2(2)
30	1	4	170, 190, 200, 230		4(2)
30	1	7	80, 80, 90, 130, 170, 170, 260	1	6(2)
30	2	1	260, 340		2(1)
30	2	1	80, 170	2	
30	6	1	190, 210, 230, 310, 180, 270		6(2)
40	1	7	130, 140, 280, 80, 220, 220, 240	2	5
50	1	6	60, 80, 120, 140, 210, 230	1	5
50	1	7	80, 90(±), 160, 160, 180, 230, 230	2	5
80	1	2	410, 370		2(2)
100	1	2	180, 240		2
150	4	1	120, 370, 160, 290		4
200	7	1	180, 210, 100, 140, 180, 210, 290		7
200	3	1	190, 280, 300		3
200	8	1	80, 100, 130, 190, 270, 270, 130, 190		8

with great caution, until they chanced upon a run, which they would then quickly mark with urine several times. Eventually the run would lead them to the entrance of a hole where they would stop with quivering tails. After a while they would enter the hole A few minutes later Telle heard squeaks and then the intruder would usually come running out of the hole, often followed by other rats. Mostly, however there was only one pursuer to each intruder. Once the intruder had abandoned the run he was left severely alone. If he was very young and remained in the pack's territory then he was usually found dead within a few days, but Telle was unable to discover any wounds or internal injuries. He assumed that these animals had died from hypoglycaemic shock. Barnett has observed the same phenomenon: when a healthy wild rat is placed in a cage with strange rats, or simply picked up in the hand, it may die from sheer fright.

Telle was unable to determine the precise boundaries of a given territory. Where strange rats are driven off one day intruders may be tolerated the next, and vice versa. There was, however, one rule of thumb he could establish: the closer he comes to the hole the greater the chance that the intruder will be driven off. This is true of many species of animals living in burrows and nests, but with rats it applies to small packs only.

Black rats defend their territory more fiercely than brown rats do. They also fight outside their runs. Generally several members of the pack join together to attack the intruder.

Telle also conducted very interesting experiments into the acceptance of intruders. Unfortunately, this work was confined to only two sets of experiments and needs to be repeated and extended. On the first occasion, Telle released an intruder among

Table 5. The acceptance and expulsion of intruders by black rats

Size of pack	No. of intruders	No. of experiments	Weight (in g.)	Result	
				expelled	not expelled
10	1	1	120	1	
10	1	3	130, 140, 180	3	
20	1	6	80, 90, 160, 170, 110, 170	6	
20	2	1	70, 130	2	
30	2	1	140, 220	1	1 (1)
30	3	1	110, 140, 180	1	2 (2)
60	2	1	180, 210		2
150	2	1	210, 265		2
220	6	1	190, 200, 140, 180, 220, 220		4

a pack of eighty animals living in a pigsty; on the second occasion he released an intruder among a pack of one hundred
animals in a large chicken house. The intruders were not driven
off. Subsequently he trapped and killed sixty-two animals in
the pigsty (in five days) and seventy-six animals in the chicken
house. Fourteen days later he again released two intruders in the
pigsty and twenty days later two intruders in the chicken house.
The intruders in the pigsty were driven out after two hours;
those in the chicken house after thirty minutes. Telle repeated
this experiment in another chicken house, inhabited by seventy-
eight rats. A stranger was added but was not chased away. After
a week, six animals were captured. A day later five rats were
introduced: three of the six captured animals and two strange
rats. None were driven off. The results of all these experiments
suggest that members of reduced packs will not behave as
members of small packs until after some time has elapsed.
Unfortunately, Telle conducted too few experiments to draw
any firm conclusions. One would like to know if this kind of
behaviour is the general rule (which means examining at least
twenty cases) and after exactly how many days a reduced pack
will start to expel intruders. Nevertheless Telle's experiments,
of which experimental psychologists would do well to take
notice, are of greater value for the understanding of rat
behaviour than all sorts of experiments conducted in animal
psychologists' mazes.

From Telle's general results I am inclined to assume that
animals in large packs do not (or hardly) know one another,
which explains why intruders are so readily accepted: they cannot be recognized as strangers. Animals in small packs, by contrast, know one another so that strangers are quickly identified.
Animals in large packs that become thinned out learn to know
one another after the lapse of some time and are then able to
recognize strangers, who are subsequently expelled.

Telle also placed strangers in an area that had previously been
cleared of all rats. There were no corpses; the strangers combined to form a pack. When Steiniger released strange rats in a
"free enclosure" two of the animals formed a pair that quickly
killed all the other rats (the female becoming a specialist in
severing the carotid artery). Alas, neither Telle nor Steiniger re-

peated these experiments, but the difference in the behaviour of the two sets of rats may be the difference between freedom and imprisonment.

PAIR FORMATION AND RANK

No pair formation has been observed among free-living rats. Sexually receptive females are followed by strings of sniffing males, each trying in turn to mount them. Barnett has described fights among males for such females, but Telle was unable to discover anything of the kind. If a female on heat is introduced into a large cage containing numerous males, fights between the males will break out after some time but chiefly because the male rats, who are highly motivated sexually by the presence of the female, try to mount one another. The males that are mounted usually try to ward off these unwanted sexual advances and a series of grim but short fights ensues as a result.

Steiniger has written at some length about the ranking order in a pack. This order becomes particularly obvious during feeding: some animals are only allowed to eat when others have finished their meal. Moreover, none but dominant males are allowed to mate with a female on heat. Steiniger further discovered that when dominant males advanced into the holes of inferior males, they ceased to be dominant. However, on one occasion he observed a fleeing male entering the hole of a male over which the fleeing animal was dominant. In that case, the inferior animal was driven out of its own hole. But all this happened in Steiniger's "free enclosure". Telle was quite unable to establish the existence of a hierarchy in a free-living pack, and when I myself studied a pack on a refuse tip, I too was quite unable to discover anything like a hierarchy, although I gained the impression that one of the old males tended to be dominant. Of course, it was quite possible that the social order was by then so well established that no fights or skirmishes were needed to preserve the social balance: everyone knew his proper place. In this pack, incidentally, there was no such thing as privileged access to food or to females by one or several males. Even in a large cage in which I kept a pack for a long time and in which one of the animals *was* dominant, the kind of situation described by Calhoun never occurred; each animal could eat whatever he wanted, and couple with a sexually receptive female whenever there was one and no other animal was already engaged in coupling with her. Nor did the dominant male always copulate first. In Calhoun's experiments, 150 adult rats were kept in an area measuring 100 square feet. The result can only be described as a behavioural sink. There is (I shall return to this point at greater length) no reason to look upon this experiment as anything but a scientific oddity, and least of all to treat it as a key to human behaviour as was done by Robert Ardrey among others. This is not only unwarranted and un-scientific—it is downright dangerous.

ABOUT THE BLACK RAT

Far less is known about the black rat than about its brown relative. Nevertheless several interesting studies have been published by Eibl-Eibesfeldt, Mrs Ewer, W. F. Pippin, J. S. Watson and C. B. Worth. Mrs Ewer's monograph on black rats is a particularly valuable contribution to our knowledge of the behaviour and habits of these animals. During a period of two years she spent almost every evening observing a group of free-living black rats near her laboratory in Accra, Ghana. It was not her original intention to work with black rats, but since food she put down for other animals was always consumed by black rats she decided to make a virtue out of necessity and to proceed to a study of these uninvited guests. She observed the rats shortly before sunset (5 p.m.) and again from 8 p.m. onwards. The rats quickly became used to her presence. She did not mark the animals but was able to identify them individually from a missing bit of ear or tail, etc. The first thing she established was that black rats are extremely good climbers; they can easily scale up vertical branches. They are also capable of running across telegraph wires, describing a wide figure eight with their tails. There was one female, part of whose tail was missing, and who consequently found it exceedingly hard to negotiate the wires. The tail also plays an important role when an animal slips, in which case it can be whipped round a branch or a wire.

Propagation was not found to be subject to seasonal fluc-tuations. There was a great deal of emigration: healthy young adults and adolescents would regularly disappear. Now young animals, more than six weeks old, were constantly attacked by older animals, and their only possible responses were to flee or to make appeasement gestures. In the first case, they were pursued and attacked time and again until they emigrated for good; in the second case they were left in peace, and allowed to remain with the group. A striking feature in the expulsion of some young rats is that some of the older females may keep picking up particular youngsters. The victims develop into peripheral members of this group, creeping up to food when it is available, snatching a morsel and running off with it to a quiet spot. And one day they, too, disappear for good. Thus there is a constant flow of emigrants, balanced by some immigration. Mrs Ewer prefers to speak of fluctuating groups rather than fixed colonies or packs because, as she points out, a population of rats is dynamic, not static. The males in a group will accept strange females without further ado; the females in the group, by contrast, will attack strange females. However, if a strange female resolutely refuses to flee, she will eventually be accepted by the rest. A strange male must defeat the most important male (there is thus a hierarchy) of the group before its presence is tolerated. However, if a female in a group is in oestrus, there is suddenly a large influx of strange males who try to mount her. These males are attacked; there are protracted fights but never any dead—at worst there are insignificant wounds, and the males are not usually driven off.

Within a given group there is a great deal of aggression. Mrs Ewer describes it carefully (I shall come back to this point in Chapter 5), and from her description it appears that aggression is not nearly as intense as it is among rats kept in cages. Most of it is confined to threats and approaches, unless a female is in oestrus. She also mentions a ranking order, that is, the existence of a definite top male among the older rats. No hierarchy can be observed among younger males which does not, of course, mean that none exists. If the number of females is small then a hierarchy appears among the females as well; if there are many females no such hierarchy occurs. In general, Mrs Ewer puts it that an animal will attack anything that is smaller than itself; females however do not hesitate to attack larger animals as well. In one week she counted thirty attacks by two females against intruders; in the same week she counted three attacks by two males against intruders. It should however be mentioned that if males attack, they continue to fight for a much longer time.

Females are very rarely attacked by males. (Much the same is true of brown rats.) Thus the top male in Mrs Ewer's group would come up regularly to the corner of the veranda to beg nuts from Mrs Ewer, Every other male that approached during this time was immediately chased away; a female, however, would be tolerated.

Unfortunately Mrs Ewer's is the only thorough study of a group of black rats. Pippin has made some interesting observations, discovering, *inter alia*, that rats which he released five miles from the place where he originally captured them found their way back home within two days. But neither his observations nor Worth's are comparable to the work of Mrs Ewer because neither made protracted studies of a group. Mrs Ewer's observations largely agree with Telle's. Her findings about the treatment of intruders and the expulsion of young adults in particular are of great importance and it would be extremely valuable if these observations could be repeated and expanded by further experiments with free-living rats.

RATS IN STRANGE PLACES

Rats can live in the strangest of places. Thus colonies of rats have been found in coal-mines, a subject on which a voluminous report has been published by the British government. At a time when horses still lived in the mines, rats and mice must have found plenty of food and warmth below ground. On islands, too, rats will usually multiply prodigiously within a fairly short time, especially when brooding sea birds provide them with a profusion of food. The rats studied by Habs and Heinz on the North Sea island of Scharhörn and by Steiniger on the North Sea island of Norderney fed largely on birds and on birds eggs—indeed Steiniger's five thousand rats had become specialist bird-catchers. They would run about as if the birds did not concern them in the least, and then suddenly pounce on their unsuspecting victims. Steiniger believes that this was an acquired skill, not a faculty based on "general instinctive foundations". Most modern ethologists would eschew this choice of phrase, but no one would blame Steiniger for it—our terminology becomes antiquated within ten years. Nevertheless I believe that Steiniger was wrong. Brown rats throughout the world have a strong predilection for birds. For that very reason I believe that their ability to learn how to catch birds is so strongly programmed genetically as to rest squarely on what Steiniger has called "general instinctive foundations". The rats on St Helena would pull roosting chickens out of the trees at the time of Napoleon's exile. Thus Las Casas made the following entry in his diary on Thursday, 27 June 1816: "We had to dispense with our breakfast; an eruption of rats that had issued from various points in the kitchen during the night had carried everything off. We are literally overrun by them. They are enormous, mischievous and very bold; it took them only a little time to pierce our walls and our floors. The duration of our meal was enough for them to penetrate the *salon*, where they were immediately attracted to the vicinity of the table. It happened more than once that we had to do battle with them after dessert, and one evening, the Emperor wishing to retire, the one in our company who tried to hand him his hat caused one of the largest of rats to jump out of it. Our grooms tried to raise some poultry, but had to give up the idea because the rats devoured everything. They went so far as to climb into the trees and to kill the fowls roosting in them." O'Meara, Napoleon's physician on St Helena, has also described a constant war against rats. It was difficult to poison them because the dead animals could not be removed from under the floorboards and between the walls and so spread an intolerable stench. And, O'Meara goes on to say: "At night, startled by their sudden appearance in my room and their scuttling across my bed, I would fling my boots, my bootjack or anything else I could seize at them, but without the least effect, so that finally I was obliged to get out of bed to chase them away." It is tempting to quote further evidence showing that rats spread across islands with incredible speed and that, as a result of the ensuing struggle for existence, they became bold enough to attack human beings. But I lack the space and I would moreover impair my plea for the rat if I recited all the horror stories that can be culled from eighteenth- and nineteenth-century travelogues.

In the trenches and on ships rats can be very great nuisances indeed. During the First World War, provisions had to be hung up from wires in the trenches lest the rats devour all the food at night. Ships are usually infested with black rats. Before the Second World War, an annual average of four thousand rats per large port was recorded in Britain after the fumigation of ships. Among these four thousand rats an average of seven were brown rats; all the others were black.

PREDATORS

Alas, very little is known about the relationship between rats and the many animals that prey upon them. One possible procedure is to analyse the contents of the crop and the droppings of birds of prey. Thus Uttendorfer has shown that buzzards, sparrow-hawks, black and red kites, goshawks, hobbies, peregrine falcons and kestrels regularly feed on rats. But rats only form a small percentage of the diet of these birds. Owls, too, make good ratters. According to Ticehurst 9.7 per cent of the food of the brown owl, and according to Leslie 6.7 per cent of the food of the long-eared owl, consists of rats. Other species of owl also feed on rats. This very fact led to the large-scale disappearance of owls and other birds of prey from the Dutch province of Zeeland at the beginning of 1973—these birds had eaten rats poisoned with coumarin. In places where there are many owls it is therefore important to use only such poisons as make rats die in their burrows.

According to Southern and Watson (1941) only 1.7 per cent of the food of the foxes investigated by them consisted of rats. No similar data are available for stone and pine martens, stoats, polecats and weasels. What is certain is that these animals do feed on rats among other animals. Otters, too, will sometimes catch rats and it is often claimed that they prey on musk-rats as well. This makes it all the more incomprehensible that human beings should have gone to such lengths in their efforts to eradicate the otter. A good stock of otters is more likely to keep rats at bay than any of the poisons we are in the habit of using against them.

4.
RAT KINGS

A UNIQUE DISCOVERY

On a cold February day during the hard winter of 1963, a Dutch farmer, P. van Nijnatten, heard loud squeals while he was standing in his farmyard in Rucphen (North Brabant). "Going in the direction of the noise," wrote A. J. Ophof in *Rat en Muis* of November 1964, "he spotted a black rat (Rattus rattus) looking out from underneath a pile of bean sticks in a barn. Unfortunately he killed the rat, but when he tried to pull it out from the pile he found that the animal would not budge. On closer investigation he discovered that its tail was tied to the tails of six other black rats. These animals, too, he killed. The barn itself was clean and tidy—the rats were found lying on the floor and not in a nest.

"In the barn there was no food. There were numerous rats in the adjoining chicken coop but all these, as far as could be ascertained, were brown rats (*Rattus norvegicus*). In the barn itself no black rats had ever been seen before, but there were black rats in the loft of the farm house, which was some twenty yards away from the barn. . . .

"The rat king consisted of seven adult rats (*Rattus rattus*) [see Fig. 9], namely five females and two males, all of the same age to judge by their size. They all looked well fed. The knot [see Fig. 10] included almost the entire tail of one animal but only the tips of the tails of the rest. So as not to destroy this rat king, the knot was not untied. As far as could be ascertained the tails had not grown together. However, they seemed dented at their point of contact with other tails, and some sections in the knot appeared swollen. The X-ray of the knot [see Fig. 11] showed some fractures, both in the tail itself and also in some of the vertebrae. Some of the fractures, at least on the original negative, showed signs of a callus formation. All these findings suggest that the tails had been knotted together for some considerable time, and that some of the caudal vertebrae may have begun to atrophy. This could have been confirmed by a histological examination but only at the cost of destroying the 'king'. A single protruding tail was necrotic, but only at the extreme tip. The knot also contained bits of straw. The nails of some of the animals were worn away.''

THE ORIGINS OF THE NAME

It seems very odd to describe a number of rats whose tails are tied together in such a way that they cannot escape as a ''rat king'', but this name has interesting historical origins. It is a mediaeval German term; the English name is a direct translation of the German *Rattenkönig*, and so probably is the French *roi des rats*. The term was originally applied, not to rats, but to one who lives well on the backs of his fellows. I think it likely that the word was used well before Luther first put it in writing in 1524, and also before Konrad Gesner defined it in his *Historia animalium* (1551-8): ''Some would have it that the rat waxes mighty in its old age and is fed by its young: this is what is called the rat king.'' Luther put it as follows: ''The archbishops have a primate above them, the primates a patriarch and, finally, there is the Pope, the king of the rats, right at the top.'' This particular usage was in keeping with his description of cardinals as ''that rabble of rats'', of monasteries as ''rats' nests'', and of the theocratic state of the Anabaptists in Munster as a ''rats' kingdom''. Later, the term came to be used for a crowned king sitting on a throne of knotted tails.

The first to depict a rat king was Johannes Sambucus, who did so in his *Emblemata*, published in Antwerp in 1564 (see Fig. 8). Curiously enough, that particular king is not reproduced in the excellent treatise by Becker and Kemper on this subject. In any case the phenomenon must have been well-known in Luther's day and little else could have symbolized the power of the Pope more graphically than a rat enthroned on the knotted tails of its conspecifics. It is not known how this particular idea first arose. G. Schellhammer mentioned it in 1691 and Bellerman, who published his *Über das bisher bezweifelte*

Daseyn des Rattenkönigs ("On the hitherto doubted existence of the rat king") in 1820, suggested that the idea went back to an Oriental myth.

In 1757, Noel Gomel was the first to include the term "rat king" in a dictionary, namely to refer to a number of rats joined together by their tails, and various other dictionaries followed suit in the eighteenth and nineteenth centuries. The German writer Jean Paul seems to have been the first to use the term to refer to an "entanglement" or "Gordian knot" and it is still used in that way, particularly in German literature.

The French term *roi des rats* is commonly thought to be a direct translation of the German *Rattenkönig*, but Poisson and Pesson, Dollfus and Hughes believe that the word was derived from *rouet* (spinning-wheel), the intertwined tails being the spokes of the wheel (they were originally referred to as *rouets des rats*). The idea may be far-fetched but it is certainly ingenious.

THE FINDS

The oldest rat king known to me is that which was depicted by Sambucus (1564). The poem accompanying the illustration speaks of a gentleman who was plagued by rodents for years and of a servant who came across seven rats with their tails tied together. The word "rat king" does not occur in this poem. The most recent rat king known to me is the king discovered in Rucphen in 1963, which was also the first Dutch rat king ever described. From 1564 to 1963, fifty-seven rat kings were discovered and described, chiefly in Germany. Of these finds, eighteen are not fully authentic. In their zeal to mention kings, all sorts of writers have included unreliable reports in their articles and monographs, just as I am now doing myself. One of these "finds" must,. moreover, be considered a deliberate falsification.

Sambucus merely said of his king that it consisted of seven rats. Later reports give more information. The five rat kings discovered in the seventeenth century are listed in Table 6.

The rat king of Danzig was first mentioned in a letter from a local professor to a colleague in Basel. This letter, discovered by

Table 6. Rat kings discovered in the seventeenth century

Date	Place	Number	Age	Authenticity	Condition	First record
20.3.1612	Dantzig	9	adult	–	alive	Reh (1926)
4.7.1683	Strasbourg	6	young	+	alive	*Mercure Galant* (1683)
1690	Kiel	14	?	+	alive	1691
1691	Weimar	?	?	–	alive	1691
8.6.1694	Krossen	15	?	+	alive	1751

Reh, states that on 20 March 1612, the writer discovered a rat king consisting of nine living, well-fed animals behind a partition in a loft. The rat king of Strasbourg was the subject of a print first published in the *Mercure Galant* in 1683. It conveys a very good picture of the king. Later prints were often highly stylized. One pamphlet included an illustration of a king accompanied by a lengthy text, according to which God sends us rat kings to remind us of our sinfulness. Apart from a detailed summary of man's various sins, the tract also contains a good description of this particular find. The text mentions two dates, 4 July and 14 July, because the calendar was changed at this time. The king was found in the cellar of a man called Würtzen, and consisted of six "strikingly large rats with their tails so intertwined and fused that they could not be separated without injury". The find was taken for public exhibition at the town hall and subsequently all but one of the animals, which managed to escape, were killed. This murder of defenceless animals inspired the writer of the tract to produce the following effusion: "May God thus scatter and thwart all wicked and poisonous attacks on this, His Christian band, lest such evil gain the upper hand and take root in the dark pits of bloodthirsty hearts."

The rat king of Kiel was discovered under the tiled kitchen floor of a prominent citizen. A squeaking had often been heard from beneath the floor and rats had been seen in a hole. In order to kill these rats boiling water was poured down the hole, whereupon four rats jumped out. The squeaking, however, continued. The tiles were then taken up and a rat was spotted that made no attempt to run away. When pulled out by the

8. Emblem of Johannes Sambucus (1564). Note the rat king.

9. The rat king of Rucphen (1963).
10. Detail of the knotted tails of the rat king of Rucphen.

maid with a poker, the animal came up but without a tail. After a second attempt "a large monster" was lifted up: fourteen adult rats with their tails joined together and squeaking loudly. The animals were immediately thrown into the privy and drowned.

Table 7. Rat kings discovered in 1700-1750.

Date	Place	Number	Authenticity	Condition	First record
1705	Keula	?	+	alive	Döring (1914)
7-1719	Rossla	9	+	alive	Linck (1727)
1719	Gödern	?	–	?	Linck (1727)
1722	Dieskau	12	+	alive	Liefmann (1723)
1722	Leipzig	many	+	dead	Liefmann (1723)
5-1722	Tambachs-hof	5	–	dead	Linck (1727)
3-1725	Dorndorf	11	+	alive	Hellern (1731)
1727	Werni-gerode	?	–	dead	Linck (1727)
12-7-1748	Grossball-hausen	18	+	alive	Goeze (1787)
26-12-1748	Langen-salza	10	+	alive	*Witt. Wochen-blatt* (1779)

Schellhammer mentions another rat king (in Weimar) but gives no further details. The rat king of Krossen was discovered in a mill. The animals were killed with boiling water, hung from an oak tree so that hundreds of people could inspect the king, and finally thrown away.

During the first half of the eighteenth century ten rat kings were reported, seven of the reports being historically reliable. Linck (1727) made engravings of the kings of Rossla and Tambachshof. These are highly stylized prints and do not convey a proper impression of the animals concerned. The Tambachshof print has caused many to doubt the authenticity of this king since the tails do not appear to be tied together. All the rats in the Rossla king were of the same size; the king of Dorndorf consisted of rats of various sizes, probably a mother with her young, and the king of Langensalza consisted of ten well-fed males. This king was discovered in a monastery, and like the kings of Leipzig and Keula was pickled in alcohol, these specimens having since unfortunately disappeared. Most of the kings discovered at the time had boiling water poured over them, after which they were disposed of. Linck described four kings in 1727. One of these was obviously authentic, but of the other three cases it is not certain whether they were genuinely discovered at the time and in the way described by Linck. The king of Wernigerode is said to have been preserved by Count Stollberg, but an inquiry to the Stollberg family has produced no further information. The rat king of Dieskau was discovered in a barrel of peas. The barrel was quickly covered, boiling water was poured in, and a king of twelve rats was later removed. This king was taken into Dresden to the Royal Natural History Collection, but unfortunately this king, too, has been lost, probably during a fire in 1849.

The rat king of Gross-Ballhausen was found in a mill by the miller, Johann Heinrich Jäger. Although it would seem that a copper engraving of this king was made—Goeze mentioned this fact—the print is no longer extant.

Of the rat kings listed in Table 8 not a single one has been preserved. Of the two rat kings of Bernburg we know from the brothers Beckmann that the first was discovered in a mill "long ago" (hence well before 1751) and that the second was found in a cellar and that it was in a desiccated state when the brothers inspected it. There were still seven of the original nine rats left. The rat king of Arnstadt was the subject of five oil paintings, of which four were probably lost during the Second World War. The extant painting ($22\frac{1}{2}''$ x $15''$) is in the Castle Museum, Arnstadt. The knot is very accurately reproduced, and so are the

Table 8. Rat kings discovered in 1750-1800

Date	Place	Number	Authenticity	Condition	First record
1751	Bernburg	11	−	?	Beckmann (1751)
1751	Bernburg	9	−	?	Beckmann (1751)
26-11-1759	Arnstadt	6	+	alive	Bellermann (1820)
1772	Erfurt	12	+	alive	Bellermann (1820)
12-1-1774	Lindenau	16	?	alive	*Witt. Wochenblatt* (1774)
1793	Wundersleben	10	+	alive	Bellermann (1820)
1793	Brunswick	7	+	alive	Bellermann (1820)

rats, which were obviously black. We know nothing about the circumstances surrounding this discovery, except that the king was found in the home of the tinsmith Georg Heinrich Schönherr, "down by the market".

Bellermann's monograph contains an illustration of the rat king of Erfurt. The knot is very stylized and does not resemble Bellermann's own description or that of others who actually saw this king. The rat king of Lindenau is discussed below. A copper engraving of this king is found in the Leipzig Local History Museum. The knots look rather artificial. The rat king of Wundersleben was discovered in a stable and the rat king of Braunschweig in a privy.

During the first half of the last century several very famous kings were discovered (see Table 9). The rat kings of Buchheim and Flein have been preserved; they are found respectively in the Altenberg Mauritianum and the Stuttgart Natural History Museum. There are several good photographs of these kings and also of a rat king of Bonn which was kept in the University

Table 9. Rat kings discovered in 1800-1850

Date	Place	Number	Authenticity	Condition	First record
1810	Ath (Belgium)	11	–	?	Goffin 1937
1810	Brunswick	7(adult)	+	alive	Meisner (1818)
1817	Frankfurt	?	–	?	Treitschke (1840)
1817	Niedersachswerfen	?	–	?	Bellermann (1820)
12-1822	Döllstadt	28(adult)	+	alive	Lenz (1835)
12-1822	Döllstadt	14(adult)	+	alive	Lenz (1835)
5-1828	Buchheim	32 (young?)	+	dead	Cuvier (1831)
5-1829	Flein	8(adult)	+	alive	Hase (1914)
10-1-1830	Frankfurt	13(adult)	–	alive	Anon. (1831)
1835	?	?	–	alive/dead	Voigt (1835)
3-1837	Zaisenhausen	12	+	alive	Kilian (1837)
1841	Bonn	6(adult)	+	?	Blasius (1857)
1843	Stammheim	6	–	alive	Thuringia (1843)
15-3-1844	Leuterhausen	7(adult)	+	alive	Kilian (1844)

Zoological Institute of that town but lost during the Second
World War. There is a drawing by Kilian of the rat king of
Zaisenhausen, but that drawing does not agree with the
description of Pastor Doll of Zaisenhausen, who was summoned
to the scene by the actual discoverer, a municipal councillor.
This king was pickled in alcohol and sent to Gmelin, the
Director of the Natural History Collection in Karlsruhe, but no
more was heard of it, Gmelin dying soon afterwards. The
second rat king of Braunschweig was probably kept in the local
museum for quite some time. Demaison saw a rat king there in
1884, but by 1963 the specimen could no longer be found.

As Table 9 shows, two kings were found simultaneously in
Döllstadt in December 1822. Both were discovered by threshers
in a hole six inches inside the main beam of a barn. Brehm
writes: "All forty-two seemed to be very hungry, and squeaked
continuously but looked perfectly healthy. All were of equal,
and moreover of such considerable, size that they must have
been born during the last spring. To judge by their colour they
were black rats." The rats were killed with threshing flails and
thrown on a dungheap after the inhabitants of Döllstadt had
been given a chance to inspect them at close quarters.

The rat kings of Düsseldorf, Keula and Lüneberg were
preserved for a long time but all disappeared during the Second
World War. A photograph of the king of Keula has disappeared
as well, but there are pictures of the other two kings (two good
ones of the king of Düsseldorf, a poor one of the king of
Lüneberg). The king of Dellfeld is kept in the Strasbourg
Zoological Museum and there is a good photograph of the king
of Courtalain. The other kings have all disappeared and no
illustrations are left.

So far, ten authentic kings have been discovered in the
twentieth century. The kings of Hamburg, Rüdersdorf and
Rucphen have been preserved and the king of Java may also still
be extant (in Buitenzorg). The king of Moers was preserved but
has disappeared and the king of Berlin has been disentangled.
With the exception of the kings of Buitenzorg (Java), Büngern,
Le Vernet and Châlon-sur-Marne, photographs are available of
all these kings.

Apart from the king of Java, which consisted of field-rats

Table 10. Rat kings discovered in 1850-1900

Date	Place	Number	Authenticity	Condition	First record
19-9-1860	Berlin	8(old)	–	?	Berl. Nachrichten (1860)
1860	Aldringen	?	–	?	Steinvorth (1884)
±1865	Zaunröden	5	–	alive	Becke (1965)
1870	Keula	?	+	?	Döring (1964)
2-2-1880	Düsseldorf	8 (young)	+	alive	Beckmann (1880)
15-4-1883	Lüneburg	8(adult)	+	alive	Lehmann (1884)
1889	Obermodern	5 or 6 (young)	+	alive	Scherdlin (1919)
1894	Vézenobres	6	–	alive	Hughes (1937)
4-1894	Dellfeld	10 (young)	+	frozen	Scherdlin (1919)
11-1899	Courtalain	7 (young)	+	alive	Oustalet (1900)

Table 11. Rat kings discovered in the twentieth century

Date	Place	Number	Authenticity	Condition	First record
8-5-1905	Hamburg	7 (young)	+	alive	Becker (1964)
1906	Le Vernet	7 (young)	+	alive	Buysson (1906)
27-1-1907	Rüdersdorf	10(adult)	+	?	v.d. Meer Mohr (1918)
5-10-1914	Moers	7 (young)	+	alive	Otto (1921)
23-3-1918	Java	10 (young)	+	alive	v.d. Meer Mohr (1918)
21-10-1937	Büngern	9	+	?	Nieuwe Rotterd. Courant (1937)
1940	Lichtenplatte	5 (young)	+	alive	Olt (1940)
2-6-1949	Berlin	3(adult)	+	alive	Orion (1952)
1951	Châlon sur Marne	4(adult)	+	alive	Giban
2-1963	Rucphen	7(adult)	+	alive	Ophof (1964)

(*Rattus brevicaudatus*), all the other kings were of black rats, at least as far as one can tell from the pictures. The tables show that only rarely is a dead rat king discovered, most kings drawing attention to themselves by their loud squealing.

DISCOVERING A KING

When a prominent Braunschweig citizen heard continuous squealing in his house for several days he had floorboards removed. Seven pitifully squeaking rats that were so weak that they could hardly move were discovered in a small hole. All their tails had been joined together so firmly and so inextricably that they could not be pulled apart and the entire king had to be lifted as one. Almost every find elicits similar reports. Thus, on 2 February 1880, a postman, C. Fischer, heard unusual squeaks that seemed to come from high up a wall near the roof. In Lichtenplatte, people heard protracted squealing in the pigsty. On 15 April 1883, a rat king was discovered in Lüneberg at the back of the house of the merchant Ohler that had made itself known by loud squeals from underneath a lavatory. Many more such stories could be reported.

When most rat kings are discovered, they are already fully formed. There was, however, one interesting exception. In Berlin on 2 June 1949, at about 5 p.m., three captured rats were thrown into a bucket. The next morning at nine o'clock Herr Otto Janack, an official in the municipal rodent extermination department, was called to inspect the bucket. Inside sat the three rats, tails entangled. Herr Janack, who had never heard of rat kings, considered it all a poor joke, the more so as it was only with great difficulty that he was able to disentangle the pretzel-shaped knot. He then took the rats to the Animal Hygiene Laboratory of the Robert Koch Institute. According to Becker and Kemper, Herr Janack was a thoroughly trustworthy person. The king must thus have been produced within a few hours. Perhaps someone had indeed played a joke on Herr Janack, although it is no laughing matter to tie three living rats together by their tails.

The rat king of Büngern was found by a servant in a star-ling's nest in the yard of a farmer called Housel, and consisted of nine rats. Unfortunately we have no further details about this king, which is not mentioned in Becker and Kemper's monograph.

EXTANT KINGS

The following kings may still be examined: the rat kings of Hamburg, Dellfeld, Buchheim, Flein, Rüdershausen and Rucphen. The rat king of Buchheim was the subject of a controversy because for many years it seemed impossible to tell of precisely how many animals it consisted. H. Grosse, the present-day curator of the Mauritianum, carefully counted the king in 1963: twenty-seven animals were still wholly preserved; five animals had partly disintegrated. This, the most famous of all kings, was discovered when the miller Steinbruck of Buchheim, near Eisenberg in Germany, had his chimney opened up. Inside he found a desiccated and hairless black king, which had probably been badly scorched.

The rat king of Dellfeld was peculiar inasmuch as three of its ten rats had teeth marks on their heads and front legs. Before this king was immersed in alcohol pieces of incisor were discovered in the gnawed legs and heads. There are two possible explanations: either the animals had bitten one another or else other rats had attacked the defenceless king. When it was found under a bale of hay the king was completely frozen, and lengths of hay were wound up in the knot.

The rat king of Flein consisted of eight males measuring about six inches from head to tail. The individuals constituting this king were not arranged in the usual circle but looked like a bunch of flowers with the tails representing the knotted stems.

The rat king of Hamburg consisted of seven rats, one of which had slipped out of the knot. The individual rats had a length of about four inches. They were probably all of the same age and looked well fed. This king is kept in the Hamburg Natural History Museum.

The Rüderhausen rat king consisted of ten rats of which seven have been preserved. The size of the animals ranged from four to six inches, and they seemed well fed. No details about the actual discovery are known, except that a timber merchant called Degenhardt presented the find to a museum in Göttingen when it was still "fresh".

Becker and Kemper, having read A. Ophof's report about the fractured tails and vertebrae and about callus formation in the rat king of Rucphen, went on to examine whether similar features were present in other kings. They discovered that such fractures were found to a greater or lesser extent in most kings, but not the calluses. Callus formation may indicate that a king was several days old when it was discovered; yet the nutritional condition of most of the animals found seems to suggest that the king was either of quite recent origin or else that it was fed by outsiders. Now a well-fed rat can easily stay alive for forty-eight hours without food. It will then be very hungry but not yet very emaciated. If it is left without fluid for a long time then it seems to shrink and may appear younger than it really is. Hence the rather vague term "good state of nutrition" does not necessarily mean that the king had recently been formed. The other possibility—that the king was fed by other rats—is something I would dismiss out of hand. On several occasions I have starved a brown rat for a whole day and then placed it in a cage connected to a second cage containing another rat with large supplies of food. The "rich" rat could easily have pushed food through the opening to the "poor" rat, had it wished to. Even when this experiment was performed with mothers and their young, the mother having the food, the young were never fed through the opening. And whenever I placed the hungry animal inside the cage of the "rich" rat, the only food it was allowed to eat were the crumbs the other had scattered about by chance. Rat kings may, of course, have procured food in similar ways—if other rats were present near by they may well have scattered enough crumbs to keep the king alive. For the rest, while brown rats can be *taught* to feed others, this is no part of their natural behavioural repertoire. Things may, of course, be different with black rats, but I consider this most unlikely.

THE KING DISPLAYED

It is usual for kings to be put on display for some time after discovery. The two kings of Döllstadt were exhibited in various places; that consisting of twenty-eight rats could be seen in an inn and that with fourteen rats was on show in the woodman's shed in which it had been found. Occasionally newly discovered kings have been paraded through towns or villages so that everyone could have a good look at them. The rat king of Leipzig was kept—from 1722 onwards—in mummified form in the private museum of Dr Petermann. I often imagine the man whom I admire more than anyone else, with the possible exception of Mozart, Johann Sebastian Bach—who lived in Leipzig from 5 May 1723—going to view this particular king during a visit to Dr Petermann's museum. Unfortunately he composed no music for the occasion. Rats portrayed in music are very rare indeed, and since there is not enough material for a chapter on the subject I shall say a few words about it here. The most important rat in music is found in Prokofiev's beautiful opera *The Love for Three Oranges*. This particular rat is an ugly transformation of the Princess Ninetta. Both Schubert and Wolf have put Goethe's poem *Der Rattenfänger* ("The Rat-Catcher") to music, Wolf turning it into a particularly lovely song. The Dutch composer Koumans set several of La Fontaine's fables very beautifully to music. Gounod's *Faust* makes use of the first lines of Goethe's rat song, and in Berlioz's *La damnation de Faust* The whole poem is set to music in a most spirited and original way. The entire work is exceptionally beautiful: I bought a record for the sake of the rat song alone, but the music so inspired me that I went out and bought all Berlioz's recorded works. What writing a short book on rats can lead one to!

But let us return to the city of Johann Sebastian Bach. In 1774 it was possible to inspect another rat king in Leipzig: the rat king of Lindenau. Anyone wishing to view it had to pay an entrance fee to the painter Johan Adam Fassauer. The *Wittenberger Wochenblatt* reported that interest was keen and that Fassauer had earned a great deal of money. He also produced a copper engraving of this king which can still be seen in the Leipzig Local History Museum. The knot of the sixteen tails looks as if it could be disentangled with a single tug. It bears little resemblance to the knots of other well-known kings. Because Fassauer was always short of money, Geyser (1858) has suggested that the painter caught these rats himself, killed them and tied their tails together. There is also a report that the rat king was found by one Christian Kaiser, a miller's assistant, on 12 January 1774, and that he handed it over to Fassauer so that the latter could make an engraving of it. When Fassauer started to earn money with the king, Kaiser demanded his specimen back and all the money as well. The end of the story is unknown.

EXPERIMENTAL KINGS

There have been several attempts to produce kings experimentally. Thus Herman Landois killed ten young brown rats in 1883 and tied their tails together. Unfortunately this pseudo-king can no longer be found in Münster, where it was kept for a long time. One thing, however, is certain: anyone who ties up the tails of dead rats (I have tried it several times) will obtain something that in no way resembles the kings found in nature: the knots are too neat, and look like the one depicted in Fassauer's engraving. His story had shown that it was lucrative to own a king, and so people began tying tails together. Küsthardt (1915) reports that many such sham kings were exhibited at fairs and similar gatherings.

In 1966, Wierts took an important step forward. He, too, tried to create artificial kings. He used white laboratory rats (albino forms, not of the black, but of the brown rats). During the first experiment he used seven young rats weighing $1\frac{1}{2}$ oz., with tails $3\frac{3}{4}$ inches long, all from the same nest. He glued together the last $\frac{1}{2}$ inch of the tails of three and four of these rats respectively, allowed the glue to dry and then placed both groups together in a cage measuring 8 x 12 x 8 inches. After a few minutes, a true rat king had been created: as the animals leapt and crawled about their tails became entangled in a proper knot. After four hours the animals, which had no doubt been squealing piteously, were released. Here, too, some of the caudal vertebrae had been fractured and the tails were found to be partly discoloured and partly atrophied.

In a second experiment the tail ends of six animals, weighing $3\frac{1}{4}$–$4\frac{1}{4}$ ounces and with tails about $4\frac{1}{2}$ inches long, were glued together. Here, too, a rat king was created within a few minutes of the glue having dried and the rats having been released in a cage. The glue was then removed under anaesthetic. When the animals came to, one was able to escape and the rest proved exceptionally intractable. The knot was eventually untied.

Though I like to repeat most experiments that I read about because then one can truly understand what is going on and come to grips with the interpretations of the original workers, this particular experiment is much too horrible for me to try. The experimental animals would have to be killed since it would be cruel to keep animals with mutilated tails alive. My courage fails me.

It is clear that rat kings are formed very quickly once the tails have been glued together by man, indicating that they can also come about quickly in nature. But this also means that, provided one knows the trick, it is easy to produce artificial kings. The question whether or not kings were made by human hands can not be settled by experiment.

However there are other arguments to show that many rat kings came about in natural ways. From the early accounts of the finds, we know that kings used to frighten people, who were far from happy with such discoveries. Moreover most people refused to touch animals that were believed to cause various diseases. Hence it seems highly unlikely that they should have touched rats for the doubtful benefit of producing a king. Moreover we know of the existence of kings among other animals with long tails. And finally, even artificial kings demand live rats, and it is far from simple to catch six or more rats all of the same age and to glue their tails together or tie them up in a knot. It is fairly easy to do this with albinos under an anaesthetic, but anaesthetics are not usually kept in mills, kitchens, privies and all the other places where kings were found.

But the most important argument in favour of the existence of genuine kings—and absolutely decisive to me—is the following: so far as we can still ascertain, all the kings discovered consisted of black rats. Now, since the number of black rats has declined greatly since 1700, one would have expected to find an increasing number of artificial brown rat kings, for the chances of catching black rats have ~~in~~*de*creased proportionately. Wiertz's experiments have shown that it is quite possible to produce a king of brown rats. Yet never has the discovery of a brown rat king been reported. And not surprisingly so: the brown rat has a much shorter and less pliable tail than the black rat, and the chances of the tails becoming naturally intertwined is thus much greater in the second species. Perhaps there are also differences in behaviour—unfortunately very little is known about that of the black rat—which increase the chance of the appearance of kings among them.

Marshall, a Leipzig professor, declared in 1903 that rat kings must be considered so many practical jokes (I for my part can see nothing funny in them), but Brehm was convinced of the natural origin of most kings. A number of my colleagues also do not believe that kings could have been produced along non-experimental paths, i.e. by other than human hands, but many change their minds after examining photographs of the actual knots. Those who do not are unable to explain convincingly why all the kings should consist exclusively of black rats. For a long time, I myself used to think that the whole thing was a fable, but after carefully examining the long tails of black rats I realized how easily they could become tied together. And having read Becker and Kemper's excellent monograph I have become convinced that rat kings are a natural phenomenon for which we still have to find a full explanation.

KINGS OF OTHER ANIMALS

On 31 December 1951, McClung discovered seven squirrels in a zoo in South Carolina sitting on the ground with their tails tied together. He caught them and took them to the animal hospital. Two females were dead, a third female was dying. The other animals were released but only after their tails had been cut off above the knot. All the animals were adults. A squirrel king had been found previously in the same zoo, and there is also a report of the discovery in 1948 of a third squirrel king of three animals.

McClung could not tell how the king he discovered had been

formed. All the kings at the zoo were found when the weather was both cold and snowing. Had the animals huddled together for warmth and was that why their tails had become entangled?

In Europe, two red squirrel kings have been discovered: in August 1921 a king consisting of five animals was found, of which one had died. After the other animals had been released their tails fell off down to a stump measuring $\frac{4}{5}$ of an inch. The second king was discovered on 20 October 1951. In this case too it consisted of five animals which were killed by the finder, Herr Dunkel, and pickled, and can still be inspected. There is also an excellent photograph of this particular king.

Reports exist of a few kings of house mice, but since there are no reliable data, photographs or specimens, I doubt whether an authentic king of such animals has ever been found.

THE ORIGINS OF RAT KINGS

Let us recall the known facts. Kings have only been found among black rats; the chances that a king will emerge are clearly connected with the length of the tails. The animals are often of the same age and, in at least half the cases in which the age was reported, the individuals were not yet adult. Very large kings are the exception; most kings consist of five to twelve animals. The rats are generally found alive; they are not particularly emaciated and most are discovered because of their loud squeals. Swellings, shrivelled tail ends, fractures of the vertebrae and breaks in the tails all suggest that the animals try desperately to free themselves. The kings have been discovered in the sort of places (behind walls, in lofts, in cellars, in barns, etc.) where black rats generally make their nests. It is therefore conceivable that, at least in a number of cases, the animals had crowded together in a nest when the king was formed. We know exactly when a few of the kings were discovered: eleven were found in the winter, eight in the spring, four in the summer and four in the autumn.

Many kings had extraneous material introduced into the knot: hay in the tails of the king of Dellfeld; straw in the kings of Hamburg, Java and Rucphen; tallow, clay and cow hair in the king of Düsseldorf; dirt in the king of Hamburg and a "pitch-like, sticky substance'' in the king of Moers.

One of the oldest theories about the origins of rat kings was advanced by Linck in 1726. Since cats are not infrequently born with knotted umbilical cords, Linck assumed that the same might happen with rats, and that when the animals later try to break the umbilical knot, their tails become intertwined. It has also been suggested that the tails become intertwined in the womb—the animals are born as kings. It is a plausible idea, the more so as the individual members of the king are usually of the same age and still young. They might (except in the case of very large kings) well belong to one and the same litter. The tail of the black rat is, however, particularly short at birth (roughly $\frac{3}{5}$ of an inch); once they are born, the tail begins to grow more quickly than the rest of the body and continues to do so until tail and body are each four inches long, whereupon the tail grows at an even rate with the body. Moreover, the tails of young brown rats are the same length as the tails of black rats. Hence, if black rats could indeed form a king in the womb, it ought to be possible for brown (or white) kings to be formed in the same way. Yet although untold numbers of white rats have been born since 1900, not a single king has ever been discovered among them. It therefore seems highly unlikely that rat kings are formed in the womb. Nor do I think that such kings would survive for even one day: a mother rat rejects any young that does not look perfectly normal.

Steiniger has also dismissed the suggestion that the tails might be glued together by the afterbirth; following birth the mother cleans the young from head to tail.

Meisner (1818) has suggested that healthy rats might tie up the tails of old and weak conspecifics to provide themselves with a good nest. Blumenbach and Richter believe that old rats huddle together for warmth and that, in the process, they crawl among one another, eventually entangling their tails. However, Goeze (1792) had already pointed out that kings usually consist of healthy, lively animals. One might laugh at Meisner's theory as one might also laugh at the whole idea that rat kings come about by natural means. But then people also laughed when they were told that sea otters crack mussels on stones they place on their stomachs while floating on their backs, or that hedgehogs transport apples by spiking them on their spines. In biology one must never laugh at any hypothesis; one should either try to support it with sound argument, or else challenge it with equally good argument—or leave well alone.

Goeze also put forward a hypothesis of his own. He believed that in the mating season the males fight for possession of the females and that their tails become entangled in the process. But very often it is young, not sexually mature, animals that are found to constitute the king. Goeze's hypothesis does not therefore fit all the cases.

An entirely different hypothesis was put forward first by Burdach, and later by Oustalet and then Steiniger: in the winter, rats huddle closely together to keep one another warm; their tails become soaked with urine and freeze together, and during attempts to free themselves the animals jump wildly about, thus creating a rat king. This strikes me as a plausible explanation, the more so as it is well known that rats' tails can indeed freeze up. It could, however, only apply during severe winters, and, as we have seen, several live rat kings have been discovered at other times of the year.

Becker and Kemper believe that kings may be produced when the animals sit closely together for grooming and are suddenly frightened (for instance, by someone pouring boiling water into their hole), in which case they all try to escape so quickly and in such confusion that their tails become entangled. In this connection Becker and Kemper mention the phenomenon of "thigmotaxis". This is a splendid word for an animal's tendency to touch things with its skin or tail. But like so many learned terms, it explains absolutely nothing. In any case it is a fact that black rats are able to wrap their tails round a branch, for instance if they start to fall out of a tree. Perhaps, therefore, a frightened rat might indeed wind its tail round the tail of a conspecific with all the consequences we have seen.

Before a rat king can be formed, all the animals must be very close to one another. Now, black rats are contact-shy animals (in contrast to brown rats which try as much as possible to sleep, etc. in a huddle), although when it is very cold they will sit closely together. A king might be formed under those circumstances. The animals also keep touching one another during play, when five or six young at a time (mostly members of one and the same litter) will tumble about. This play is a continuous and complex motion taking place at a very great speed and it seems quite possible that a rat king could be formed in the course of such behaviour, certainly when some sticky substance is present in the nest or in the playing area. Now, since rats pick up all sorts of things it is not inconceivable that they should bring back some sticky matter into their nest.

To make quite sure one would have to study the behaviour of black rats in the laboratory. Black rats are much more beautiful than brown rats, so why not? Perhaps we should then at last be able to solve the mystery of rat king formation.

MULISCH ON RAT KINGS

In the Netherlands very little has been written on the subject of rat kings. One instructive contribution was published by van der Meer Mohr in *Tropische Natuur* in 1918 (No. 9, pp. 113-15) and there is the excellent article by A. J. Ophof in *Rat en Muis* of November 1964, which deals chiefly with the rat king of Rucphen. These journals do not, however, enjoy a mass circulation in the Netherlands, a country in which knowledge of rat kings is nevertheless fairly widespread. Most people derive this knowledge from Harry Mulisch, who devoted a whole page to the subject in his *Bericht aan de rattenkoning* ("Report on the rat king"). It is a great pity that he also perpetuated so many errors. He claims that rat kings are dead when they are discovered, which is very rarely the case. He says that the caudal vertebrae are fused. This has never been found. He also claims that kings are kept alive by other rats. There is no proof of this claim. He says that it is impossible to tell whether rat kings are formed before or after birth. It is almost certain that they are not formed before birth. His remark, "We know of rat kings consisting of twenty-seven rats", is very misleading because it suggests that we know of many such rat kings. Mulisch must have heard of the king of Buchheim, which was for a long time wrongly believed to have consisted of twenty-seven rats. As it is, there is no single known rat king consisting of that number. Mulisch's claim that "Luther already referred to the subject" is misleading. Luther called the Pope a rat king but did not use the term in its modern sense. At the beginning of his piece Mulisch says: "A rat is an extraordinarily intelligent animal, as countless laboratory proofs have shown." Laboratory proofs show that rats have good memories, a good sense of smell and good exploratory powers. But such tests tell us very little about their intelligence. Certain maze experiments might suggest that rats are possessed of insight, but the interpretation of these experiments is particularly difficult. For brown rats, which live in burrows in the wild, must needs be able to find their way quickly in complicated tunnel systems, and they do so with the help of inherited skills, about which very little indeed is known. In traversing a complicated maze a rat must always make choices: to the left, to the right or straight ahead. If the rat becomes used to a maze in which, for instance, it has to start by turning right, and is then transferred to a maze built as a mirror image of the first, it will again begin by turning right, but usually notice its mistake, return to the first crossroads and turn left instead. Then continuing, it will, after perhaps committing another one or two mistakes, keep making the "mirror" choice, turning left where it used to turn left. In short, it would seem that, after its first mistaken attempt, the animal comes to realize the true nature of the transformation. But is this real insight? There are problems that are apparently very much simpler and that rats nevertheless fail to solve. These are problems that never occur in the rat's natural environment. What strikes us as being insight may very well be instinct, though it is very difficult to draw a clear distinction between the two. Those who work with mazes may all too soon find themselves in a maze of their own making.

If by intelligence we understand the display of insight—and what else can it be?—then rats are intelligent creatures. But if one compares this sort of intelligence with, for instance, the intelligence of cats, it is quite certain that cats are infinitely more intelligent than rats.

Mulisch also says: "One should never compare a man with an

animal, for such comparisons are *always* offensive to the animal. A rat, for instance, is harmful and spreads plague, but that is as nothing in comparison with what man is and spreads.'' And with that I am in complete agreement.

11. X-ray photograph of the rat king of Rucphen.

5.
THE BEHAVIOUR OF RATS

There has been a surprisingly large number of publications on such aspects of the behaviour of white laboratory rats as exploration, feeding and mating, but very much less is known about grooming, nest-building and the ontogenesis (i.e. the development of the youthful phase) of behaviour. Moreover, the *structure* of aggressive behaviour has hardly been investigated at all. Tolman stated as late as 1958 that rats show few if any complex social behaviour patterns. Munn dealt with the same problem in a handbook (1950), summarizing all past research work on rats in a chapter entitled "Abnormal and social behaviour". He began the very short chapter on social behaviour (10 pages in a book of 474 pages) with the remark: "Relatively little research has been done on social behaviour in rats primarily because rats are not especially influenced by each other's action." Further on he added: "One observes in rats none of the complexities of behavioural interaction found in higher mammals." Such comments leave me completely astounded. Did these men never look at two rats in a cage? Or did they wake up sleeping rats in the middle of the day, and rats, moreover, that had always lived in small cages without ever being picked up and thus encouraged to move about more actively? That must surely be the reason they saw so little. Nevertheless I find it almost incredible that, in half a century of research with rats, no one should ever have looked at their social behaviour to contradict the idle musings of Tolman and Munn. It all makes one wonder about the calibre of experimental psychologists during the first half of our century.

In fact, the social behaviour of rats is unusually complex and richly structured. This can only be seen if the behaviour is carefully observed, preferably in the evening hours when rats are most active, and if the research is not confined to rats that have been kept in otherwise empty cages since early youth.

An ethologist about to study the behaviour of animals begins by naming the various actions he observes. As a result, he sets up a sort of behavioural catalogue, an ethogram as it is called. Now as soon as one wants to draw up an ethogram of the behaviour of rats, unsuspected difficulties arise. In rats, and the same is true of many other animals, different forms of behaviour do not make way suddenly for others as happens with, say, the three-spined stickleback. That fish fans water or secretes a sticky fluid, digs pits, turns on its side, or performs a zig-zag dance. All these forms of behaviour begin and end quite suddenly. With a stop-watch, it is possible to time the duration of each form of behaviour exactly. Moreover, everyone can quickly learn to distinguish between the various actions, and most investigators who work with sticklebacks are agreed on the naming of the various types of behaviour. In short, all of them have adopted the same ethogram. But with rats there are no such abrupt transitions, only forms of behaviour that shade into one another, with the possible exception of grooming, feeding and copulation. Every student of rats has his own ethogram, usually based on the list drawn up by Grant. Students tend to translate their doubts in their own powers of observation into ever more subtle distinctions, and so the list of behavioural patterns grows until it becomes too unwieldy for most practical purposes. In this chapter I shall present an ethogram of the behaviour of rats such as has been used by most ethologists during the past ten years, with, of course, additions of my own. I shall also make a few remarks about the structure of rat behaviour, for which purpose I shall have to use a few ethological terms. These terms will recur in the chapter entitled "Reward and punishment", and I think it best to introduce them by way of an example.

ETHOLOGY

Ethology is the study of animal behaviour. It is by no means the same as animal psychology, as is often claimed; it is a fundamental biological science which emerged in the 1930s when Lorenz, followed by Tinbergen, began to observe the behaviour of animals in nature.

Behaviour can be defined as the sum total of all the active movements an animal performs. However, this definition has to be taken very broadly. Sleeping, too, is a form of behaviour, and so is breathing. With animals it is impossible to study behaviour by asking questions or by introspection; one can only observe. The novice investigator, however, is inclined to interpret behaviour by introspection. If an animal creeps away he says the animal is afraid or the animal is shy. Even accepted ethological terms still contain subjective elements, for instance when ethologists describe the call of a certain goose as a "triumph cry". British ethologists in particular have tried to circumvent this problem by describing behavioural patterns in as neutral language as possible. The animal goes to the left, the bird lifts its wing, etc.

To dissuade students of psychology from interpreting the behaviour of animals by means of subjective feelings I usually begin my ethology lectures to them be presenting them with a hydra (fresh-water polyp). It is inordinately difficult to associate this animal, which looks like a vase of flowers, with such feelings as rage, fear, shyness, melancholy, etc. It is not that an ethologist denies the existence of such feelings; he cannot, however, observe them, and hence feels that he had best keep quiet about them.

A hydra clings with the base of its vase to the bottom of a ditch or a puddle or to a plant, while waving its tentacles about. The tentacles contain cells with a coiled-up mechanism, often provided with a barb. Comparable cells contain structures ensuring propagation. A hydra moves by a series of somersaults. It attaches itself to an object with a tentacle, the foot of the vase is released and the animal then somersaults over its tentacles, and so on. When the cells presiding over the capture of prey are in operation, the cells ensuring propagation cannot work, and *vice versa*. There is thus a kind of mutual inhibition.

When do the prey-capturing cells work? Two conditions must be satisfied. In the first place the hydra must be hungry, and in the second place it must be stimulated, first mechanically and then chemically. The mechanical stimulus appears when suitable prey swims up against the hydra. If the prey introduces a reduced form of glutathione into the water then the hydra

12. Upright posture (left); reconnoitring (right).
13. Stretching.

14. A young rat washing itself (front view).
15. A young rat washing itself (side view).
16. Licking of front paws during washing.

shoots off its barbs. Only an adequate stimulus will produce an adequate reaction; in that case we speak of a key stimulus. The hydra must also be hungry. A satisfied hydra does not react to key stimuli. This difference in response to one and the same stimulus, which depends exclusively on the inner state of the hydra, is a difference in motivation. A hydra may or may not be motivated to eat. This response can be measured by providing suitable prey and counting the number of barbs released, which, incidentally, is not an easy task. There is, however, another method of measuring motivation: the hungrier the hydra the longer its tentacles and body. Thus the ethologist has at least two ways of measuring motivation indirectly. With other animals there is no such simple method of measuring motivation.

If a hydra is stimulated mechanically, for instance by being touched with a glass rod, it will not shoot out its barbs but draw in its tentacles. If we now keep touching the animal with the glass rod it will, at a given moment, cease to pull in its tentacles. We then speak of habituation to the stimulus.

One may ask: why does a hydra take food? Two answers are possible. One can say: it takes food in order to stay alive. This defines the *function* of taking food. One can also say that a hydra takes food because, due to a lack of certain chemical substances, the tension in the cells containing the barbs becomes so great, and is increased further by the presence of glutathione, that the barbs are shot out. (This, incidentally, is only one possible explanation of what happens; it is not quite certain whether the barb-releasing mechanism works in this way.) This explains why the barbs are shot out (the *cause* of the attack). Ethological studies are always studies of either the function or else of the causes of behaviour. Functional studies usually produce problems concerning the evolution of behaviour, the continued existence of the species and so on; causal studies usually involve problems of a biochemical and physiological nature.   .

I hope that, by the example of the hydra, I have been able to clarify certain ethological concepts to the reader's satisfaction.

NON-SOCIAL BEHAVIOUR

Exploration

As soon as a rat finds itself in an unfamiliar environment, it will begin to explore that environment intensively. This exploration consists of a number of actions that are easily distinguished. The rat always begins by *reconnoitring* (see Fig. 12). It runs gingerly about, sniffing cautiously at all the objects it encounters on the ground. During this process the rat will often adopt an *upright posture* (see Fig. 12), raising its forelimbs and inhaling vigorously.

When a rat approaches an unfamiliar object (and this is particularly true of wild brown rats), or an object with which it has had bad experiences in the past, it will, at a certain distance from the object, stretch both front legs out towards the object, its hind legs remaining in place. The animal thus appears thin and elongated, and may remain fully stretched in this way contemplating the object for some time. The object may well be another rat. This *stretching* (see Fig. 13) can take place at various distances from the object, the distance depending, *inter alia*, on the strength of the unpleasant experience. No matter how unpleasant the experience may have been, however, the rat will always run in the direction of the spot where, for instance, its whiskers have been singed; the animal wants both to stay at a fixed distance from, but also to run towards, the spot where it has just experienced something unpleasant. This is a good example of so-called ambivalent behaviour.

During exploration the animal may also *climb* (see Fig. 7) while continuing with the intensive sniffing that is so characteristic of an exploring rat.

Mazes encourage exploration in rats: intensive exploration familiarizes them with the maze; rewards in the form of food at the end of the maze are of subsidiary importance. A rat that can negotiate a maze quickly has in fact been taught to stop exploring, a fact that was not fully appreciated for a long time. Rats used to be selected for their ability to pass quickly through mazes, the so-called "maze-bright" being picked out and the so-called "maze-dulls" being rejected. But when workers first took the trouble to observe maze-dulls during their progress

17. Eating (left) and sleeping (right).
18. Lying down.

19. Grooming of rival and pressing down with paws.
20. Grooming and pressing down of rival during play.

through the maze, it emerged that these animals devoted much more time to exploration: they were more lively and more inquisitive than the "brights". The result was a spate of interesting papers which J. H. M. Vossens has summarized in his thesis on the explorative behaviour of rats.

Grooming

Once a rat has investigated its new surroundings thoroughly, it will usually begin to groom itself. Almost without exception it will begin with *washing* (Figs. 14, 15 and 16), moving its front paws across its head, first close to the mouth, then over an ever-wider area until finally it moves the paws from behind the ears right across the face, bending the ears forward with every stroke. Often the animal will lick both its front paws during this process. The paws move simultaneously across the face. Although the whole manoeuvre looks very much like the washing of a cat, there is a characteristic difference between the two: a rat uses both front paws while a cat uses just one. Washing is always followed by *grooming*, except when washing serves as a displacement activity, for instance during a fight. When grooming itself, the animal will nibble at its flanks and hind legs, generally starting at the front end of the body and ending with the tail.

Rats will also *scratch* themselves with one of their hind legs, concentrating chiefly on the flanks but sometimes scratching the insides of the ear as well. After scratching, they will often *lick* the toes of the *hind leg*. Scratching may occur independently of washing and grooming but can also occur immediately after these types of behaviour. Under the influence of various drugs (for instance scopolamine) washing and grooming may be suppressed while scratching increases, possibly because the drug produces itching, but observations have shown that scratching, washing and grooming are independent of one another in other situations as well.

Many ethologists believe that grooming only occurs when the animal is not strongly motivated to do something else and/or if there are no external stimuli to trigger off other forms of behaviour. Rats, they believe, have a tendency to groom themselves continuously, which is often inhibited by the stronger inclination to investigate, to eat, to fight, to couple, etc. As soon as a rat does none of these things it will begin to groom itself. Grooming thus represents a vacuum in the general behavioural repertoire which is quickly filled and which ensures that the animal cares for its skin at moments when no more vital responses are elicited. It also ensures that vital responses are not inhibited by grooming.

Feeding
We distinguish between the following behaviour patterns:

Eating (see Fig. 17). The animals generally sit on their hind legs and hold the food between their front legs. In a sense rodents, just like men, have their hands free and it is strange that no super-rodent has evolved in the way that a super-ape evolved from apes walking bipedally.

Drinking. Rats lap up water with their tongues.

Scattering food. This happens very often. Thus when Barnett offered his rats large pieces of liver, he observed that they pulled the liver in various directions. This does not support the claim that rats deliberately share their food. On the other hand, there is the oft-repeated story of two rats collaborating in the transport of an egg. One rat is said to lie on its back, the other to place the egg on to the stomach of the prostrate partner, who holds it between its legs while the first drags the partner away by the tail. La Fontaine wrote a fable on the subject but the story is much older still. The journal *De levende natuur* ("Living Nature") has cited several instances of this type of behaviour, and I have seen a poor photograph of it.

Two rats can be taught to behave in this way, though not very easily, but does this mean that such behaviour occurs in nature too? You have only to speak to a group of country women, apparently thoroughly reliable, to hear some say that they have seen it with their own eyes.

Now we know that all rats enjoy eating eggs. A rat is, however, able to carry an egg unaided (provided it is not too large) by biting two small holes in the shell, hooking its teeth into the shell, and raising its head. The egg may then be transported to a safe place. No assistance is required. In many other situations, too, in which collaboration would be a great advantage, I have never seen anything of the sort, unless the rats were taught it first. The question of whether or not free-living rats do convey eggs in the manner described by La Fontaine is a difficult one to answer. It is not impossible, but it is highly unlikely.

Hoarding. One cannot say that rats hoard food in the manner of squirrels or jays. The mere fact that they lay in supplies of perishables stamps them as amateurs in this field. Barnett, in his *The Rat*, has published a bibliography of the vast literature on the alleged hoarding activities of laboratory rats.

Covering food. Rats cover their food as soon as they are in a position to do so. If they are given a newspaper they will chew it to pieces and use the paper to cover any food and water they have been handed in an open tray until both are hidden from view. As regards their diet, brown rats will eat anything although they prefer animal products. Black rats are often said to be vegetarians, but Mrs Ewer's black rats consumed mice, insects, frogs and snakes. Brown rats, too, will flourish on a vegetarian diet, e.g. seeds, as Drummond, among others, has been able to show. Brown rats can also fish, and so, according to Mrs Ewer, can black rats. They will "filter" the water in a small stream through their paws and then catch the fish. Brown rats also dig for mussels and eat the contents of the shells. Such behaviour has often been observed, especially along the British coast.

However, although rats are omnivorous, they will only eat familiar food, treating strange substances with the greatest suspicion. Even when wild rats are very hungry they may often leave strange food untouched for as much as a week. Galef has shown that this type of behaviour is acquired in early youth; the composition of the mother's milk teaches the young something about what the mother has eaten. Later, the young follow the mother and observe what solid food she picks up and what she passes by. Thirdly, the young prefer to take their food from the mouth or front paw of the mother or another older rat, who will usually allow them to do so. This is not just a friendly action but one that teaches the young what food their elders are eating.

Nesting

Nest-building is associated with such forms of behaviour as *gnawing* and *dragging*, which also occur in other situations. Rats will chew paper, straw, wood shavings and all sorts of other objects into fairly small fragments, and work them into nests in the shape of small bowls. Although brown rats do not arrange their nests very tidily, each successive nest is more nearly perfect than the last—nest-building improves with experience, although even those rats that have never handled nesting material will, when offered it for the first time, turn it into a fairly good nest.

If a nest is removed shortly before young are due to be born, or if the mother is not offered nesting material when she is pregnant, she will eat the young as soon as she has given birth. Nest-building is also associated with *digging*. A rat digs by using both front legs alternately. A pile of sand is built up in this way and is then shovelled aside with both hind legs.

Other forms of non-social behaviour

There are a number of behaviour patterns that cannot easily be fitted into a particular complex. Thus a rat may *take fright* while eating or grooming, but this response is not part of either type of behaviour. If a rat is frightened, it may slowly *move its head from side to side*, probably so as to increase its field of vision. If the head moves very quickly, we call it *head-shaking*.

A strange behaviour pattern is *vibration*. It can appear during washing and eating: the two front legs suddenly start to vibrate very quickly just before the animal begins to wash or to take up a pellet of food.

Finally there are such more or less involuntary motions as *shivering*, *defaecating* (see Fig. 23), *urinating* (if this happens during exploration we call it *marking*) and sneezing.

Rest

A brown rat can *sit still*, or *lie* on its stomach or on its side, sometimes fully stretched out (see Fig. 18) but more usually in

21. Evading.
22. Catalepsy in the defensive sideways posture (left) and attending (right).

23. Dominance and submission. The submissive animal is defecating. The dominant animal may be biting.
24. Biting during dominance and submission. Note the arched bodies of both animals.

a rolled-up position. The animal may yawn and/or stretch, usually arching its back and raising its head. When *sleeping*, the animal lies rolled up, slightly on one side, head resting on the front legs which will then be touching the hind legs. The eyes are generally closed.

SOCIAL BEHAVIOUR

Aggressive behaviour
Social behaviour can be distinguished into aggressive and sexual behaviour and play. Moreover, the behaviour of the mother towards her young and the behaviour of the young towards their mother can also be described as social behaviour.

The aggressive behaviour of rats is very complex indeed. In rats, aggression is dependent on age and on sex. Older male rats fight more frequently than older female rats, while younger rats play more often than they fight. There are systematic differences in the structure of the fighting behaviour of males and females, but no one has taken the trouble to analyse these differences. In what follows I shall be describing the aggressive behaviour of male animals. To a very large extent, the structure of the aggressive behaviour of female rats is the same, but there are differences in detail.

Very little is known about the cause of aggression—or about the function of aggression in rats. In any case, male rats fight when intruders enter their territory, or when they are sexually motivated (for instance by the presence of a female in oestrus) and hence try to mount one another. I have never been able to observe rats fighting for possession of a sexually receptive female. It has often been claimed but never demonstrated that rats respond aggressively to conspecifics with a strange scent. My own limited experience makes me think that, if they live in small groups, they recognize each other individually and are thus in a position to identify and attack strangers as strangers. In some experiments, rats were marked with a strange scent and, it is claimed, subsequently attacked by the rest. When I myself tried to repeat these experiments, I was quite unable to discover any such responses.

During fights, three types of aggressive behaviour can be distinguished, though the first type cannot properly be qualified as true aggression. The following actions are involved:

Attending: The rat sits quietly while looking or pointing its head at the other animal without making any attempt to approach (see Figs. 22 and 35).

Approaching: the rat runs in the direction of the rival. If it runs towards the tail of the rival we speak of *following* (see Fig. 40). Following is almost certainly a sexually motivated action, since it always gives rise to sniffing at the anal region.

Sniffing: When a rat has approached its rival it can, and often will, sniff at him for a long time. The rival will also sniff at his opponent. During these actions one of the animals may also *press the rival down* with its paws. This action is almost invariably followed by *grooming of rival* (see Fig. 19): the rat which has placed its paws on the other will now nibble and gently bite the rival's coat. Aggressive grooming is quite different from self-grooming: there is more licking and nibbling, and the nibbling may be a limited form of biting. Moreover, in contradistinction to self-grooming, the animal keeps its snout in one and the same position for longer periods. After aggressive grooming the encounter is quite frequently terminated and both animals go their separate ways. This happens chiefly when the groomed animal sits hunched up and with closed eyes. The behaviour of the groomed animal is called *crouching*. The sequence of behaviour patterns just described is observed fairly frequently, even among playing animals (Fig. 20). However, during play, approaching is almost invariably associated with *leaping up*. The sequence is repeated several times by two animals that are kept together long enough. The groomer may be groomed a moment later, and so on.

Quite different is a second form of approaching. It also begins with attending but ends with biting. Often the rival is approached at full pelt (there is a great deal to be said for distinguishing between ten different types of running, as

25. Nosing.
26. Nosing, subsequent stage. Cf. Fig. 27.

27. Nosing has given way to the offensive upright posture. The animal on the right has begun to adopt the defensive upright posture.

Timmermanns has done) and pressed down so quickly that it is thrown over. One might make a clear distinction between the two actions but the second is probably a direct consequence of the intensity of the first. In many cases, however, the animal is not thrown over, but simply lifts one front paw and closes one eye on the side on which it is being approached. We refer to this type of behaviour as *evasion* (see Fig. 21). If the other animal nevertheless continues to press the rival down, the latter will usually adopt a *defensive sideways posture*, i.e. roll over so that only one front and one hind leg touch the ground (see Fig. 22). If further force is exerted against it, it may adopt a submissive posture (see Fig. 24), i.e. lie on its back with all four paws off the ground. The attacker will then stand over the prostrate animal, generally at right angles to its longitudinal axis (*dominance*; see Fig. 23) and may start to groom the rival. This often happens if those involved are young, but in older animals the dominant male (or female) will often bite the other (see Fig. 24). The fact that *biting* is involved can be deduced from the attitude of the animal (maximum curvature) (see Fig. 24) and the *squeals* of the submissive animal. All these actions can follow one another in very quick succession; indeed a whole series of further actions can be interposed between the first approach and the final submission. What happens can only be analysed with the help of a video-recording played back in slow motion. Because of the tremendous force with which the animals can jump at each other, they cease to be complete masters of their own actions. This type of fighting is best described by the term *clinch* (see Fig. 28). A clinch consists of a series of offensive and submissive postures on the part of both animals, the alternating postures possibly reflecting the speed with which the animals roll over. In any case, the final result is one dominant and one submissive animal. The dominant animal bites the flank of the other, often getting a firm grip on the rival while bracing its hind legs against it, and pulling it up and down several times in succession. The animals may also *box* in this situation, that is, hit out with both front paws, but this only happens when they adopt an *offensive upright posture*, i.e. when the dominant animal does not stand athwart the rival and when the longitudinal axes of both coincide (see Fig. 30).

It may happen that the submissive animal remains in this position for some time after the dominant animal has departed. This is known as *catalepsy* (see Fig. 22). In other attitudes, too, an animal may *freeze*. All these are defensive types of behaviour, as are the defensive sideways posture and crouching.

During the fighting we have just described there is a great deal of biting leading to wounds in the flanks or on the back. On only one occasion have I seen one of the animals being bitten in the head. Though the wounds are not deep and generally heal very quickly, most observers who see this type of fighting for the first time appear quite horrified.

When the animals have fought for some time without any clear victor emerging they will approach each other once again. Now, however, they will run very carefully towards each other and a series of new behaviour patterns emerges. Such patterns, incidentally, also emerge during clinches, albeit in a very accelerated form as we know from a film played back in slow motion. The careful approach is followed by *nosing* (see Fig. 25), during which the animals place the tips of their noses together and sniff very intensely. Nosing often goes hand in hand with stretching. As the animals come closer to each other, they may gradually stand up to face each other, front paws on each other's shoulders or against each other's front paws. This attitude, which is very characteristic of rats and of many other rodents as well, is called the *offensive upright posture* (see Fig. 27). Students can often tell which of the two animals will submit in the end: its neck will be slightly more stretched, and in this case we speak of a *defensive upright posture*.

As we said earlier, animals in the upright posture may begin to box. Often they will retain the posture for a long time, frequently squeaking and generally also emitting low, hoarse, guttural sounds. They may gradually *flag* until their front paws touch the ground. Then each may go its own way, but more often they will rise up again and return to the offensive upright posture, bobbing up and down for a long time. The record is held by two of my old males who, on 28 February 1973, faced each other for 53 minutes and 12 seconds while gently bobbing up and down. Sometimes it will appear as if one of the animals is trying to bring its head close to that of the other, very

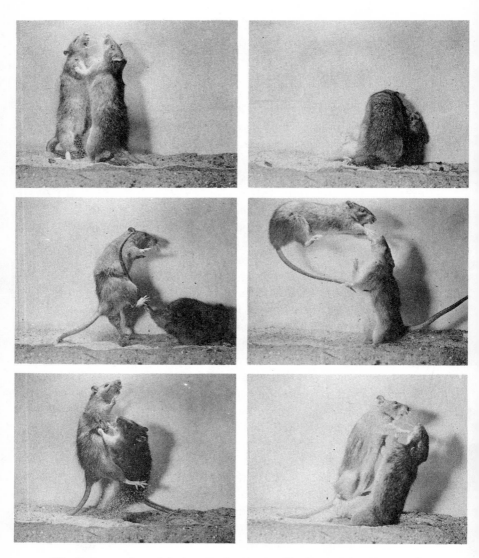

28. Postures adopted during clinch. Above: offensive upright posture (left)
and dominance/submission (right). Centre: leaping and kicking (left)
and leaping at the rival (right). Below: leaping (left and right).

29. Offensive upright posture leading to dominance and submission (Fig. 30).

30. Dominance and submission, the animals adopting quite a different posture from that depicted in Figs. 23 and 24. In this posture, there is a great deal of boxing.

occasionally biting, thus forcing him into a submissive posture. More often, however, the animals will move their heads in such a way as to keep their noses in line. As a result, the two rats look like mirror images of each other, as if one were pressing his nose against the glass while moving his head. The offensive upright posture is also adopted after the administration of certain drugs and has been described as "bizarre behaviour" by one pharmacologist. This description of what is, in fact, a common fighting posture shows how unfortunate it is that so many people who work with rats know nothing about the normal behaviour of these animals.

The offensive upright posture is usually adopted after fights, during which the two animals have bitten each other but from which no clear victor has emerged. The result is a more timid kind of approach. In ethological terms we might put it that the animals are motivated both to act aggressively and also to flee. The offensive upright posture is thus an ambivalent attitude, as we may also gather from the form of approach that often precedes it. One animal sits still, looks at the other, its front legs slightly raised off the ground. The hair of the rival stands on end and, as it approaches, it moves the rear of its trunk more quickly forward than the front. As a consequence, its body becomes arched. Finally, it stops moving its front legs but keeps running with its hind legs, turning in a circle. At the same time, two other things happen: because the back is arched, the head droops down slightly; moreover, the front leg and hind leg closest to the rival are raised, so that the animal adopts a slanting posture. Sometimes the arched back is squeezed against the rival, sometimes the moving animal will force the other into submission by pressing its concave side forward, and often this posture, which I have called the *sideways approach*, will be followed by the offensive upright posture. It may also happen that the rival will slowly *flag* into a submissive posture without having been touched (see Fig. 30). The immobile rival may also perform an action by which it can avert the threat for a moment: as the attacker approaches it can suddenly turn through an angle of 180 degrees, thus averting its head. I have called this action *circling*, though only a semi-circle is described in fact. Circling also occurs in other approach situations, but in the one I have just described it may occur several times in succession. After circling, the opponent is unable to attack and must make another sideways approach, this time from the opposite side. Circling is thus very effective but can do no more than postpone the ultimate attack.

A sideways approach is truly frightening to watch, and will have to be analysed more thoroughly than it has been so far. Words alone cannot describe it; it has to be seen to be believed, and even then it is hard to understand. To gather its full significance, one must actually fight with a rat. I have done so a few times; using certain movements of the hand one can easily imitate the sideways approach. What struck me above all was the tension of the rat when it jumped at my hand. But that, too, is difficult to put into words. Those who want to follow my example should be warned that the bites are painful but not very deep.

When one of the animals is defeated—which may happen after a single bite, it will usually rise from its submissive posture and for the rest remain perfectly still. In particular, it will have ceased to move its nose up and down incessantly, and its two front legs will be raised off the ground. The animal will adopt a more slanting position (see Fig. 35) and follow every movement of the victor who now runs about in the area in which the fight has just taken place without apparent concern. The victor behaves as if he were completely alone. The attitude of the loser has never yet been described but is nevertheless very characteristic. One might call it *crushed*, a word that, though anthropomorphic, reflects the attitude better, perhaps, than any other term. Often the crushed animal will carefully begin to explore, in which case the victor will quickly force it into submission with a new sideways approach. If the crushed animal remains stock still, the victor will approach it from time to time and *crawl*, head first, *under* the loser, usually at right angles to its motionless body (see Fig. 36). On one occasion, I saw the victor approaching the rival repeatedly with the obvious intention of crawling under him, but desisting in the end, perhaps because the beaten animal began to squeal pitifully as the victor approached. On the basis of this observation, I have come to think that this type of behaviour must be of essential importance in

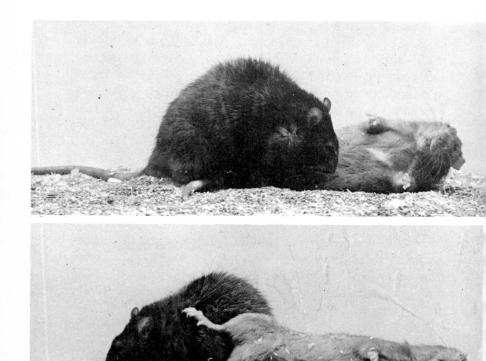

31. The posture depicted in Fig. 30 has given way to the defensive sideways posture and biting.

32. Sequel to posture depicted in Fig. 31. The victor has withdrawn and the submissive animal has become frozen in the defensive sideways posture.

33. A moment during the sideways approach. The approaching animal (left) has arched its entire body.

34. A later moment during the sideways approach. The approached animal (right) has begun to adopt a submissive posture.

aggressive situations—a victor must first crawl under its rival before he can proceed to a full investigation of the latter, and hence to complete victory. Barnett, who has described this type of behaviour, claims that it is the defeated animal which does the crawling; though this is what one might expect, it is not what happens in nature: it is the victor who crawls under the loser. It is a pity that Barnett, who is so sparing with detailed descriptions of his observations, but burdens the reader with weighty reflections, should have misinterpreted one of the few things that he did in fact observe. In general, everything Barnett has to say about aggressive behaviour in his *The Rat* is either mistaken or imperfect.

The action of crawling under the rival is not easily interpreted. The loser permits it and generally rises (or allows himself to be raised) so that the victor can proceed unimpeded. I find it hard to say why. The same behaviour also occurs in less aggressive situations, but in that case one of the two animals is usually washing itself (see Fig. 37). Alternatively, one of the animals may often, even before the fight, *climb over* the back of its rival (see Fig. 38).

The crushed animal will often emit a deep sigh, during which its whole body seems to tremble slightly. An animal can, incidentally, avoid or postpone defeat provided there is space enough: it can *flee* or *back away* from the attacker.

During, and above all after, fights, one can often hear the *chattering of teeth*. According to Barnett, this is the first act of the aggressor. I have never observed it. Mrs Ewer has described the chattering of teeth among black rats, but in these animals such behaviour occurs almost exclusively after they have been frightened. This would seem to be the case in brown rats as well; frightened brown rats do indeed chatter their teeth quite frequently.

Another type of sound can be made with the tail. House mice, in particular, will often whip their tails to and fro so quickly that a sort of rattle can be heard (*tail rattling*). I have twice observed this type of behaviour in rats, and in both cases it was performed by an animal that had been attacked shortly before.

Apart from audible sounds, rats also emit ultrasonic squeaks

both during and after fights. There are very brief impulses last-
ing from three to sixty-five milliseconds (50 kHz), or very long
impulses with a duration of 2,400 milliseconds (25 kHz). According to Sales, the short sounds accompany a very high degree
of aggression. They are also emitted by (often very aggressive)
suckling females if an intruder approaches their young. The
long sounds are emitted by rats who have lost a fight and accompany marked submissive behaviour. Sales includes cowering
among the various types of submissive behaviour. During this
type of behaviour, too, protracted sounds are emitted by the
rat, probably to prevent renewed attacks by the rival. These
sounds are also said to coincide with the deep sighs we have just
mentioned.

Audible squeaks also have a signalling function: they often
prevent aggressive acts, but not invariably so. An animal can
ward off an attack by sitting stock still and simply screaming
every time the rival approaches. As soon as it shrinks away from
the other or tries to run away (often by taking great leaps) it has
lost the fight, i.e. it will generally be attacked and bitten. Quite
a different way of warding off an attack is *kicking* (see Fig. 39),
i.e. delivering strong blows with both hind legs. This happens
above all when the other animal follows and sniffs at the anal
region. It would thus seem to be a response to an unwanted
sexual approach. Non-receptive females will also kick when they
are approached by a male. Mrs Ewer has described a number of
aggressive acts by her black rats that I have never observed
among my brown rats. There is, for instance, the *push-off* in
which the opponent is pushed away with both front legs. Brown
rats, for their part, may throw up a heap of earth between themselves and an opponent, moving the front legs alternately with
great determination—in so far as that term can be applied to
animals. Perhaps this type of behaviour is comparable to the
push-off. In addition, Mrs Ewer describes a form of behaviour
during fighting that, according to her, has the function of
appeasement: ‘‘An animal approaches a superior with hind-
quarters low, neck extended and ears usually laid back and
attempts to bring his mouth in contact with that of the
superior, aiming particularly at the sides and angle of the jaw’’.
Sounds are emitted in the process and these are more like a

35. A defeated animal (left) with its rival (right).
36. Crawling under the rival.

37. Crawling under during washing.
38. Climbing over during play.

protracted scream than the usual squeal. This behaviour is said
to prevent or to stop attacks. I have never observed this type of
behaviour among brown rats. Her remaining descriptions of the
aggressive behaviour of the black rat agree very well with what
we know of the brown rat. She gives a fairly detailed description
of rats leaping over one another in confusion, a form of
behaviour that various rodents employ in highly aggressive
situations. Thus if a number of brown rats, males and females,
are suddenly brought together, two males will usually begin to
fight with each other. The females will observe the fight with
stretched bodies but without participating in it. But if one of
the fighting animals leaps away, it may sometimes happen that
all the animals, females included, will start leaping over one
another, touching each other only by accident.

If brown rats defend themselves against a predator, (for
instance a cat) or a human being, they will first stand up (offensive upright posture) and then suddenly leap up, as may also
happen in an ordinary fight between two rats. This method is
particularly effective against cats—in one of his books
Leyhausen has reproduced a photograph of a rat jumping at a
cat, and one can see clearly that the cat is edging away. According to Leyhausen, cats rarely carry off brown rats.

Stories about leaping rats are, in any case, no fables, though
it is not true that rats aim particularly for the throats of human
beings: they will leap up, bite, and hang on by their teeth
wherever they land.

In view of the complicated nature of the motions and the speed
with which fights are fought, an analysis of aggressive behaviour
is far from simple. A video-recorder is indispensable, not least
because it is impossible to observe the behaviour of two animals
simultaneously, and because such observations are a *sine qua
non* of any serious analysis. When analysing aggressive
behaviour, the so-called time-sample method simply will not
do. In that method, one chooses a unit of time (for instance, a
minute) and records what happens during it. That method
cannot possibly lead to the analysis of a complete sequence of
behaviour patterns.

Apart from the analysis of such sequences, which will indeed throw much light on the aggressive behaviour of rats, we can also learn a great deal from analyses of the distances involved in a fight. At what distance from the approaching animal did the attacked animal begin to evade or to flag? What distances were covered during leaps or fights, and so on? We can also learn a great deal about behaviour with the help of drugs. This study is still at its beginnings but promises well. In particular, because certain types of behaviour are eliminated or intensified by drugs, this method may throw a great deal of light on the structure of behaviour. Much more important still, however, is the accurate description of the various behaviour patterns. My own descriptions are, without doubt, incomplete—anyone who observes rats carefully will be able to add to, and improve upon, them. In addition, we need accurate descriptions of the aggressive behaviour of other rodents, for only then can we hope to arrive at a uniform nomenclature for all the actions performed by the various genera constituting this order.

Sexual behaviour

It is particularly when a female is in oestrus—once every four or five days in the laboratory—that she will be followed by all the males in her cage. After following, a male will sniff at her anal region. If the female is not in oestrus, she may kick the pursuer with one of her hind legs, but more usually she will keep running while raising her abdomen slightly and also arching her back and tail (see Fig. 40). The male may try to mount the female, even if no real copulation ensues. If the female is in oestrus the situation is quite different. She will then take the initiative, running towards the male, sniffing at him, and then running off in a highly characteristic manner, with a cautious prancing or skipping movement. She will always run a few inches at a time and then sit down with raised head and quivering ears. In fact, the ears of females in oestrus will begin to quiver as soon as the head is touched anywhere between the ears, thus providing laboratory staff with a reliable method of picking out receptive females in a cage.

If the male does not follow her, the receptive female will approach him again, and again skip away. If he now follows and mounts her, she will raise her head higher still and her ears will

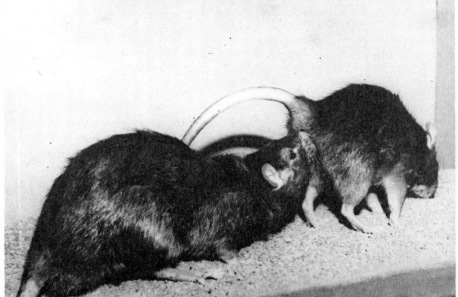

39. Kicking.
40. A male sniffing at the anal region of a female.

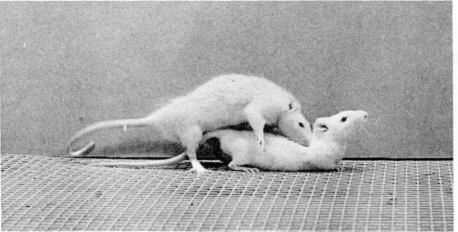

41. A male licking its penis after mounting.
42. A male mounting a female. Note the lordosis posture of the female.

begin to quiver violently until, at a given moment, they are laid back flat. Her back is now arched, the abdomen raised high. This posture is called *lordosis* (Figs. 42, 43). The male can now introduce his penis, but does not do so during the first mounting. There is a phase of *mounting* and *dismounting* during which *pilo-erection* is fairly common. After dismounting, the male invariably *licks* his *penis* (see Fig. 41). Eventually, the male will mount again and this time introduce his penis— *intromission*—without ejaculating. The whole process is repeated, the male dismounting and licking his penis several times. There follows mounting with intromission and *ejaculation* (see Fig. 43). The male closes his eyes and clings to the female with his teeth and produces a number of quick rhythmic movements. Then he dismounts and licks his penis, while she remains sitting quietly. Soon afterwards, he mounts her again. This process, too, is repeated several times. Sometimes there is a soft squealing.

If two females, one of whom is in oestrus, are placed together in a cage, they may couple for long periods. The behaviour of the mounting female then resembles that of the male in almost every detail. Thus the anal region is licked after every mounting as if it were a penis. The other female also behaves as if she were repeatedly mounted by a male.

Two males can also couple with each other, often intensively and for a long time, but I have never seen a male adopt the lordosis posture (though he will adopt something that strongly resembles it). In this situation the roles can be, and usually are, continually reversed. The mounted male may, after a few couplings, begin to mount the other. After prolonged sexual deprivation, two males will couple with each other in all sorts of ways: as soon as they touch each other in whatever position there is pilo-erection and probably ejaculation as well. This type of coupling, too, is invariably followed by licking of the penis. Both males will ejaculate and both will also lick their anal regions. I have observed the most peculiar positions, positions that are not listed in even the most progressive works on sexual behaviour. What all these animals try to do, I believe, is to bring their penis into contact with the skin of the partner. Once they have succeeded, they produce a few quick rhythmical movements of the kind associated with normal coupling, whereupon each animal will lick his penis.

On at least one occasion, working with a large cage, I observed a male coupling exclusively with another male and ignoring the many females present. It is usually said that animals only display homosexual behaviour in the absence of members of the opposite sex. In rats, on at least one occasion, this was not the case. Working with a species of barbel, K. Kortmulder similarly discovered a male who made sexual advances exclusively to another male, although females were present in the aquarium.

Play

The play of young rats consists mainly of the following sequence: attending, approaching (with much jumping), touching with paws, grooming, knocking over, and dominance/submission with boxing and further grooming. What is most marked is the quick alternation of dominant and submissive postures. Each of the two playing animals can, after having been in the dominant position for one moment, lie down in the submissive position during the next, become dominant again, lie down in submission again, and so on.

The untrained observer is quite unable to follow this pattern. The various types of behaviour succeed one another very quickly—60 to 100 types of behaviour per minute is the rule rather than the exception.

Sometimes sexual behaviour patterns also occur among young rats, but such behaviour is very rare compared with aggressive patterns.

I myself have conducted a number of experiments with rats reared in isolation. As adults, these animals evince less sexual and less aggressive behaviour than animals that have been allowed to play with other youngsters. Nevertheless, their behaviour does not change as radically as that of young apes or monkeys that have grown up without playmates.

In nature, litters usually follow one another in quick succession, for the mother is receptive again on the day she gives birth to her young. As a result, we have interactions not only between the young of one litter but also between the young of

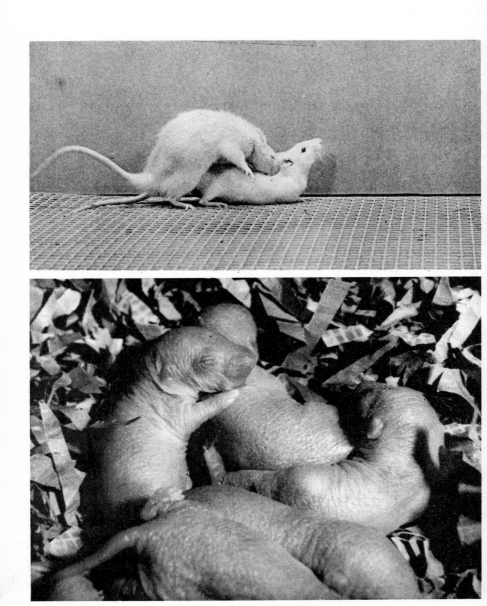

43. Successful mating with the female in the lordosis posture.
44. A nest with young rats, two days old.

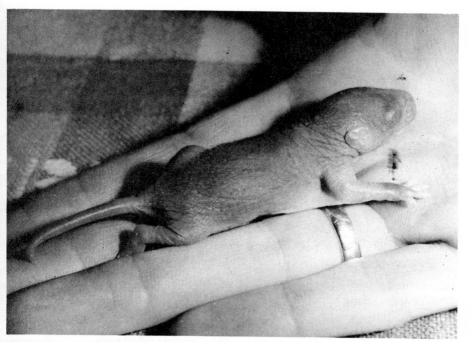

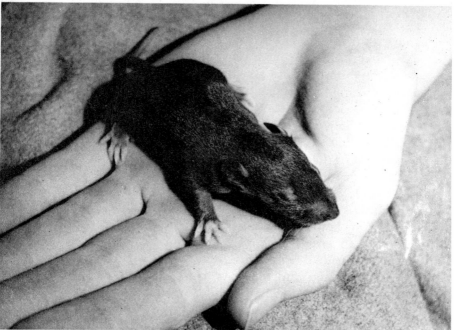

45. A newly-born rat. The tail is relatively short and stiff. It is almost impossible for these tails to become entangled in the womb.
46. A twelve-day-old rat.

various litters. It would be going too far to call the older
youngsters "nursemaids" as Jane Goodall felt free to do with
chimpanzees and jackals, but at the very least the older animals
are very closely involved in the rearing of their younger brothers
and sisters. They lick the anal region of the youngsters clean;
they bring the young back to the nest if they have strayed too far
afield, and so on. In the laboratory, as far as I know, successive
litters are not usually left together with the mother, so that in
all these respects there is quite a big difference between rats that
grow up in the wild and rats that are reared in cages.

THE ONTOGENESIS OF BEHAVIOUR

Three months after birth, a rat is an adult, that is, capable of
propagation. The history of behaviour patterns, however,
begins during the first days of life. In the laboratory, young rats
are usually weaned twenty-one days after birth, but in the wild
young rats will continue to take milk from their mothers for as
long as they are allowed to do so. Rats are born blind and hairless; within two days they are covered with a kind of filmy down
and within five days they have hair. The eyes open fourteen
days after birth. At about this time the young animals are
incredibly beautiful and exceptionally lively. Anyone who has
ever been afraid of rats ought to watch them at play at this stage
of their development—he will find it a most touching spectacle.

Immediately after birth the young rats do little but sleep and
drink. In their sleep, they will adopt all sorts of attitudes, for
instance lying flat on their backs. When sucking, they cling
firmly to the mother's teat, so much so that if the mother leaves
the nest she will drag her young along with her.

Because it is known that young pigs have their own, fixed
nipples which they defend most vigorously, K. Bonatz proceeded to an investigation of the relationship between young
rats and individual teats. This proved a laborious task because
all the young rats and the teats had to be marked, and the
mother kept licking the marks away. On the basis of the weight
increases of each of the young after each feed Bonatz was able to
calculate the milk yield: a litter will drink a total of 0.28-0.31
ounces of milk a day. There are twelve teats, though some

females have a greater or smaller number. The twelve teats lie in two groups of three times two near the front legs and hind legs respectively. The central teats are preferred; the teats closest to the front and hind legs are the least favoured. There is some individual preference for a given teat, but this does not mean that each of the young has its own teat. Bonatz was unable to discover any differences in milk yield between the various teats.

As early as their second day, the young will begin to groom themselves between sleeping and drinking; they do not actually touch their heads with their paws but simply go through the associated motions. On the third day, they begin to lick their front legs and in another two days they wash themselves properly. Scratching, too, appears on the second day and is again no more than a tentative gesture—proper scratching is not observed before the tenth day, and on the eighteenth day the animal is able to scratch its ear with its hind leg. Full grooming appears on the thirteenth day, when the young begin to groom both themselves and each other, though in the second case the animals still confine themselves to the ears.

After the tenth day the young begin to crawl out of the nest. They are always brought back by the mother who seizes them by the scruff of the neck with her teeth. The young may also leave the nest before their tenth day: the mother will often drag them out while they are clinging to her teats. Sometimes they will let go of the teats or be pulled off by some obstacle while the mother runs on. As soon as this happens, the young begin to emit ultrasonic squeaks (35–55 kHz). (Incidentally, all the young rodents studied so far act in this way.) The mother hears the sounds and runs in the direction from which they are being emitted. Experiments with sounds on tape recorders have shown that it is by the sounds alone that the mother succeeds in discovering her stray youngsters. Smell and other sense stimuli are not needed. It is the lack of warmth above all that makes the young emit these sounds. Later, when their eyes are open, they will keep leaving the nest of their own accord. At first the mother will drag them back but, after some time, she will give up the unequal struggle, for while she is busy fetching one youngster the others keep leaving the nest. During this period they will sniff intensively at solid food, but it usually takes a very long time before they will pick it up or nibble at it. They begin to sit up on their hind legs on the sixteenth day and to stand on their hind legs on the eighteenth day.

Anyone who picks up a wild young rat and returns it to the nest will find that the mother will immediately round on and kill the young one, possibly because of the strange scent. Healthy laboratory rats often show a similar reaction, though the youngster is not usually killed in the process.

From the moment they are born, young rats engage in social interactions. They push one another away while suckling, they climb over one another, and try to climb right into the middle of the litter (more warmth?) so that there is often a great deal of movement. This behaviour is gradually converted into play, with many aggressive elements. On the fifteenth day, the full repertoire of playful behaviour we have described can usually be observed.

Unfortunately very little work has been done on the ontogenesis of rat behaviour and yet such work alone can provide the essential insights needed by research workers concerned with rats at whatever stages of their development.

6.
REWARD AND PUNISHMENT

Learning is one of the most fascinating aspects of rat behaviour, and its study has involved important contributions from psychologists and ethologists alike, their respective findings complementing each other most satisfactorily.

Most of the original work was done in mazes, which have been greatly simplified over the years. The interpretation of the results, by contrast, has become increasingly complex. Until the 1950s most workers employed highly complicated systems of passages, as witness the accounts of Tolman and Barnett. Now it is more usual to use T-mazes, the simplest type imaginable.

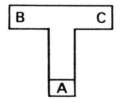

The rat is inserted at the base A, which contains a grid that can be electrified. Once it has received a shock, the animal will make straight for either B or C. If it runs first to B and is returned to the base, it may soon afterwards (say, five minutes later) make for C. This only happens if the interval between the first and second insertions is not very long. After a third insertion, there is a good chance that the rat will again make for B. This behaviour of rats has been called "spontaneous alternating", and it involves some recall of past experience. This explains why alternating has been the starting point of a most interesting study of the memory of rats.

Learning, however, is much easier to study in cages with a bar or lever, a food tray, a grid floor and a light. If a rat is inserted into this type of cage, it can, for instance, be taught to press the lever in order to obtain a pellet of food. Usually the rat is also taught that it must only press the lever when the light is on. This is done by giving the animal an electric shock whenever it presses the lever while the light is off.

How does a rat learn to press the lever? As soon as it is inserted in the cage, the so-called Skinner box, it will begin to explore the cage intensively, sniffing at the lever and by pure chance touching it with its head or front paw. As soon as that happens food appears in the tray. Sooner or later, the rat is bound to touch the lever again. Food appears again, the animal eats it, explores the tray and the lever, pressing it once more by chance, and so on. Invariably there comes a moment, often fairly soon after the rat has been put in the Skinner box (at most after 30 minutes) when the observer can see that the rat has suddenly grasped the connection between pressing the lever and the appearance of the food. From that moment on, the animal will keep pressing the lever at frequent intervals.

Anyone observing the behaviour of rats (and also of pigeons, starlings, mice and sparrows) in this situation will, if the animal is properly trained, be struck by the terrible monotony of the performance. The animal does nothing but press the lever, pick up the food, eat it, press the lever, pick up the food, eat it, and press again. This is a most unnatural way of behaving, since when a wild animal goes in search of food it will also perform a host of quite different actions, such as looking round, grooming its fur or cleaning its feathers and interacting socially.

Even if a rat is only rewarded for pressing the lever while the light is on, most of its activities while the light is off will also be directed at the lever. The animal will sniff at it, carefully place its paw on it or gently nibble at it. Sometimes where will be brief washing movements, as also occur during fights. These brief washing movements (displacement activities) betray a conflict between the desire to press the lever and fear of the resulting shock, just as, during fights, there is a conflict between the wish to attack and the fear of attacking.

Why the monotony of the Skinner box? After thirty years of experiments people have come to realize that the answer to this question has very important implications. Before I look at them in detail, I must first make a few general comments about rats in Skinner boxes. In so doing it is unavoidable that I should use a number of technical terms which I shall try to clarify as best I can.

TERMINOLOGY

A hydra reacts to the presence of whatever prey comes into contact with its tentacles or stimulates it chemically, by firing barbs at it. These so-called key stimuli are said to give rise to appropriate reactions. We speak of specific stimuli and specific responses. The word "specific" may be omitted whenever the animal has learned to react to an arbitrary stimulus (for instance seeing the lever) with an arbitrary response (pressing the lever). The learning process thus couples a stimulus (S) to a response (R), but only if R is followed by a reward. Instead of the word "reward" we can also use the term "reinforcement". An animal can also be punished after a response, in which case ethologists speak of negative reinforcement. They speak of positive reinforcement whenever a reward is involved. This type of learning is called instrumental or operant conditioning. According to Skinner, the term "operant" emphasizes the fact that behaviour operates upon the environment in order to produce an effect. The term is used both as an adjective and as a noun. The pressing of the lever is an operant. Learning theory, incidentally, provides an excellent illustration of the ability of certain scientists to use mystifying terms about what are in fact very simple matters. All of us are familiar with such operant conditioning as the training of animals in a circus. In what follows I shall simply refer to operant conditioning as training.

Operant conditioning is often contrasted with classical conditioning, i.e. with the type of learning in which a conditioned reflex is formed. This type of conditioning was studied by Pavlov. He allowed a dog to see and smell food (the stimulus). The dog produced a reflex response: it began to salivate. If now a bell was sounded the moment the food appeared, and if this process was repeated several times, the dog began to salivate whenever the bell was sounded. In that case we speak of a conditioned reflex.

If, during training, the reward is suddenly withdrawn, the animal will continue to press the lever for a long time, but the frequency, which increases soon after the withdrawal of the reward, will gradually decrease and the intervals between successive pressings will grow longer. This process is known as the extinction of the response. People have wondered whether this is an active or a passive process, i.e. whether the animal "unlearns" to press the lever because it is no longer rewarded with food (negative reinforcement?) or whether the impulse gradually ebbs away. In all probability, the extinction of a response is another learning process, but I shall not enter into the matter in further detail.

Once an animal has learned to press a lever so as to obtain food, one can begin to reward it at irregular intervals. If the sequence of intermittent rewards is regular, i.e. if five rewards in succession are followed by one omission, the animal will quickly learn to press the lever six times in succession. To avoid this, a kind of mechanical die is fitted behind the Skinner box and is thrown up with lightning speed every time the rat presses the lever. If, for instance, the die comes up showing a three, the animal is rewarded, but if any of the other five numbers comes up it is not rewarded. The result is a random reward of one in six. There are a number of ways of constructing an automatic device of this type, i.e. a device for rewarding the animal once in six tries at random (but also once in two, three or four times). In that case we speak of a variable ration (VR). In our example, VR = 1/6.)

What is the significance of rewarding an animal in this random way? In the first place, a rat is less quickly satiated if it obtains food no more than once every few times it presses the lever. In the second place, intermittent rewards help to build up a resistance to the extinction of the response; i.e. encourage the animal to persevere.

THE NATURE OF THE REWARD

We can define reward simply in terms of response in-tensification. Whenever an animal keeps pressing a lever more and more quickly to attain a particular objective, it may be said to have received a reward. It is, incidentally, of the utmost importance that the reward be given immediately after the pressing of the lever—if there is any delay, training is almost impossible. How does the rat come to treat as a reward what happens to it immediately after pressing the lever? The answer seems quite simple. The rat is hungry. It presses the lever in order to obtain some food. It presses the lever in order to fulfil a need. This is the theory of drive reduction. In the 1950s, however, a series of publications appeared which showed that drive reduction is an improbable explanation of the nature of rewards. Thus Sheffield, followed by Miller, demonstrated that hungry rats will start pressing the lever more quickly if they are given a sweet substance, although that substance does not still their hunger. The sweet taste might have acted as a reinforcer, but even if the substance was introduced directly into the stomach of an animal that had just pressed the lever, the animal began to press the lever more frequently, although not, admittedly, with the same frequency as after licking up the sweet substance. The difference may have been due to its inability to perform the action of licking.

There is yet another experiment to show that rewards are not always associated with drive reduction. Sheffield found that male rats pressed the lever more frequently if they were allowed to mount a female immediately afterwards. They were dragged off the female before they could ejaculate, so as to increase the sexual stimulation even further. Their need increased rather than decreased. It was also discovered (and I have seen this myself) that satiated rats will continue to press the lever. At that moment, no need has to be fulfilled. What then is left to reduce? Moreover, if, during training, rats are offered food in another tray from which they can take it directly without having to press a lever, they will often prefer the food they can obtain by pressing the lever. Very hungry rats, by contrast, cannot be taught to press the lever: they are so excited by the appearance of food when they first press the lever that they cannot be trained.

Rats will press the lever more frequently if the cage is lit up or darkened immediately afterwards. If there is a great deal of noise, they can be trained to press the lever for a moment of quiet. Conversely, if all is perfectly still they can be taught to press the lever for a moment of noise. From these and a number of similar experiments one might conclude that every change in the environment can serve as a reinforcer. According to McCall, the magnitude of the change influences the frequency of the lever depression in a positive sense. However, Sackett has shown that such discoveries are closely connected with the way in which the animals have grown up. In any case, sense stimulation, provided it is not too pleasant, works as a reinforcer.

The problem of the nature of rewards was given a new dimension by the experiments of Olds and Milner, who fitted electrodes to certain parts of a rat's brain. The animal was placed in a Skinner box, and as soon as it pressed the lever the brain was stimulated by the electrodes. It appeared very quickly that stimulation of a number of areas of the brain produced such pleasant sensations that the animal began to press the lever as often as it could. The brain centres concerned are often referred to as pleasure centres, but we cannot, of course, tell what precisely the rat experiences. In any case, stimulation of certain parts of the brain has a reinforcing effect and a very strong one at that, for the rats will press the lever with unusual frequency. The brain also contains centres on which electrode stimulation acts as a form of punishment. The rat will do everything it can to avoid such stimulation. Finally, there are also some neutral centres.

Valenstein placed rats into a spacious cage with ample food and with conspecifics, and stimulated such centres as had previously elicited high-frequency lever-pressing activity in the Skinner box. He observed such behaviour as eating, drinking, digging, fighting and mounting. After stimulation of those brain centres on which an electric shock has a punishing effect he observed flight and associated forms of behaviour. All in all, therefore, he was able to record all those forms of behaviour we met in the last chapter.

Does this take us any further? If an animal is hungry and sees food, that centre in its brain will be activated which regulates the ingestion of food. If a rat is not hungry and nevertheless has food before it, it will eat the food after electric stimulation of that particular centre. If the animal is given the chance of stimulating that centre itself, it will consider that chance as a reward. Vossen has accordingly defined reward as follows: "Reinforcement should be defined in terms of the activation of the brain centres involved in the production of species-specific responses." By "species-specific responses" Vossen was referring to what I would describe as "natural behaviour". One might call Vossen's definition satisfactory, but one might say with Hinde: "The mode of action of such an unnatural source of stimulation is far from clear." All that Vossen says, in fact, is that we know no more about rewards than that they are somehow linked to the natural behaviour of an animal. This vague formulation may be exemplified with the help of some cases of learning in nature.

LEARNING IN NATURE

Though Skinner's well-known *The Behaviour of Organisms* deals with learning in rats, the title suggests that what laws Skinner discovered in this study must needs apply to all organisms. More explicitly, Skinner contends that all behaviour rests on conditioning, a view that has persuaded ethologists to study learning in nature. I shall only give one example of this type of study, but it is a very telling one.

Squirrels learn to open nuts more and more efficiently from earliest youth—every nut they open helps to improve their technique. Here we might thus have a simple case of conditioning: the squirrel opens the nut for the sake of its content, i.e. the reward. However, when Eibl-Eibesfeldt provided young, inexperienced squirrels with empty nut shells that had been glued together, he made a strange discovery. "Once again," according to Sevenster, "the squirrels learned to open the empty shells with increasing skill. Since nuts are an important source of food for wild squirrels, this reaction seems to be sensible: the young animal must continue to improve its technique even when the first nuts appear to be empty." The reader may wonder what precisely the reward is in that situation. Here we have the same problem as we have just mentioned in connection with rats: the nature of the reward cannot be defined.

Nevertheless we are entitled to say that these squirrels receive some kind of reward, albeit in a special sense. By learning to open even empty nuts ever more efficiently, they increase their chances of remaining alive very considerably—something that might be called a reward. In that case reward is defined as the increase in the chances of an individual's survival.

While nuts are available, a squirrel that can open nuts has an advantage over another squirrel that cannot open nuts. However, it is impossible to tell in advance whether or not the nuts will always be available. In nature, if a situation is such that an animal, by learning something, becomes better adapted to its environment, then it is important that this animal should learn that something as quickly and as well as possible, regardless of whether or not the situation is going to change. Thus it is of great importance to a rat placed in a Skinner box to adapt itself as quickly as possible to the new situation. It can do so by learning to press the lever as efficiently as possible, not only to obtain food but also to obtain a continual stream of sensory stimuli. With such stimuli the rat provides itself with an enriched environment and hence increases its chances of survival.

The connection between natural behaviour and learning is clear: natural behaviour is always so organized as to preserve the individual and ensure the propagation of his kind. The same is true of learning: the individual learns what tends to help him adapt himself to changed circumstances, so that learning, too, contributes to his preservation and hence to his possible propagation. All this has very important implications which will be discussed in greater detail in the section entitled "Actions".

If an animal is always given the same stimulus, then, at a given moment, the natural response to the stimulus will disappear. We call this habituation. Why then is there no habituation to the stimulus "seeing the lever" in the Skinner box? Is there a difference between natural behaviour and learning?

HABITUATION

Habituation may be considered a method of adaptation to such environmental stimuli as require no further response. If a hydra keeps drawing in its tentacles after mechanical stimulation, the animal can no longer devote sufficient time to the capture of prey. It is thus of importance that it become habituated to recurring mechanical stimuli which pose no threat to its safety. Just like learning, therefore, habituation, which may also be considered a form of learning, is a process by which the individual adapts himself to given environmental conditions. As a consequence, an animal that has been adequately habituated to a stimulus cannot be easily trained to perform actions that do not recur after habituation. From various investigations, and the training of rats in particular, it has appeared that it is very difficult to train an animal to perform an action to which it has become habituated in the period preceding the training. This, too, would indicate that habituation and learning help an animal to adapt itself to the environment in the best possible way and that once the adaptation through habituation has been achieved the new relationship is not easily disturbed by further learning.

In the Skinner box, too, there is habituation. The rat placed inside the box performs a number of actions: it explores, grooms itself, rests, sits, etc., at least if it is not rewarded after pressing the lever. All these actions disappear as the rat is trained to press the lever to ever better effect. In the Skinner box, there is habituation to all stimuli but one: seeing the lever. No wonder that it is so difficult to distract a hungry rat when it is busy obtaining its food by pressing the lever. Why—and this is an important question—does the rat not become habituated to the stimulus: seeing the lever? To some extent, this is due to the monotony of which we have spoken earlier. But the question cannot be answered so easily. There is one fairly fashionable explanation for the absence of habituation to stimuli shortly before a reward. It is based on the fact that habituation goes hand in hand with the formation of a conditioned reflex. In that case habituation would be the creation of a conditioned reflex between a brain centre producing inhibition and a stimulus to which the animal normally reacts but to which it stops reacting after the emergence of the inhibition. The conditioned reflex is created fairly quickly but there is a very brief labile phase in which rewards interfere with and prevent the formation of this reflex. This means that no habituation will occur in the case of those stimuli that immediately precede the reward. Only with these stimuli is the formation of the conditioned reflex still in the labile phase. This would also mean that if the reward were given, not immediately after the depression of the lever but some time later, it would be almost impossible to train the rat. And this, as I said earlier, is indeed the case. The reward must come immediately after the response.

This is a highly theoretical explanation for the lack of habituation to the stimulus "seeing the lever" in the Skinner box. There is some evidence for the existence of such a labile phase. Vossen has discussed the whole matter at length in a lecture entitled "Comparative and psychophysiological investigation of learning processes".

ACTIONS

As we said earlier, if an animal ceases to perform an action after habituation it is very difficult to train it to perform that action again because, through habituation, it has become adapted to its environment and such adaptation cannot be easily undone. Natural forms of behaviour may also be considered so many adaptations to an animal's environment. It follows that one cannot train an animal to act against its natural inclinations. To take just one example: a rat that has been defeated must keep very still lest its rival attack it again. It follows that it will be impossible, in this situation, to train the rat to mount a female or even to press a lever. Such training would tend to disturb the animal's natural adaptation to its environment. No one has ever tried to condition rats in such situations; nevertheless, it has been stated categorically that the frequency with which any action is performed increases if the animal is rewarded immediately after the action. Thus Teitelbaum explained in 1966 that he had chosen almost any action from the animal's repertoire and rewarded it with food, water or something else for which the animal would perform it. Rats were made to press a lever and a pigeon to peck at a key, but both might equally well have been made to dance round their cages. In every learning situation, Teitelbaum asserts, the stimulus, the response and the reward are completely arbitrary and interchangeable. Not a single one of these factors has an inbuilt, fixed connection with any other. Such sweeping statements have also been made by other writers. All of them have worked with two animals, the pigeon and the rat. They used conditioning to produce a few actions more frequently than they normally appear. They have not yet looked at thousands and thousands of other species, have not yet rewarded millions of actions. Why then such confidence? It is the less justified because, in 1966, it had already been discovered that there were actions to which Teitelbaum's comments did not apply.

Let me give just one example of the limitations of learning. Since no such work has ever been done with rats, I shall use the stickleback as my illustration. A stickleback which has a nest can be trained to swim through a ring provided it is shown a sexually receptive female for ten seconds. The female is hidden behind a panel that is withdrawn the moment the male swims through the ring. This glimpse of the female is so satisfactory a reward that the stickleback learns quickly to swim through the ring as often as it can: after a fairly short period of training it will pass through the ring an average of two to three times a minute. If, however, the same stickleback must bite a rod to obtain the same reward it will bite the rod no more than once every four minutes, and some sticklebacks cannot be taught to bite the rod at all—they die after a period of horrible tremors. Not that they cannot bite the rod; indeed if another male appears behind the panel, they may bite the rod more often than two or three times a minute. However, when the animal must perform an aggressive action (biting) in order to be rewarded sexually, his learning ability is severely impaired, as this elegant experiment by Sevenster proves very clearly.

I expect that similar limitations in learning behaviour will also be discovered in rats, for instance if aggressive or defensive actions are chosen for rewards.

If learning is indeed an adaptation to change in the circumstances of an individual's life, then this is just what we

47. An eighteen-day-old rat.
48. Social interaction of young rats during a feed.

49. Young rats retain their excreta. The mother licks the anal region at periodic intervals which encourages the young to defecate. The droppings are removed by the mother immediately.

50. Brown rats are contact animals, and try to lie as close together as possible, especially when sleeping. The animals depicted are youngsters.

would expect to happen. A man in a very hot environment adapts himself by sweating. A sprinter adapts himself by producing a higher pulse rate. A beer drinker adapts himself by flushing out his kidneys. All these so-called involuntary actions are bound to be susceptible to conditioning, provided of course that such conditioning ensures better adaptation to the circumstances.

Miller and his collaborators have performed a number of delicate experiments with rats to show that the pulse rate changes under the influence of the learning process. They first located those brain centres that rats like to stimulate by pressing a lever. Next they injected these rats with curare, the well-known arrow poison of the South American Indians. Curare paralyses the voluntary and the respiratory organs, so that the rats had to be given artificial respiration. An electrode was now inserted in the brain of the experimental animals and 50 per cent of these animals were rewarded with brain stimulation whenever their pulse rate increased. The other 50 per cent were rewarded when their pulse rate decreased. The pulse rate of the first group increased by 20 per cent, that of the second group decreased by 20 per cent. In the second experiment, the rats were further expected to distinguish between two different light stimuli. Only in the wake of one of the two stimuli were they "allowed" to increase their pulse rate. This, too, they managed to do. Miller and his collaborators showed further that the blood pressure, the production of urine, and the contraction of the stomach and the intestines can all be influenced by training.

For a long time it was believed that so-called classical conditioning involved the involuntary muscles and that training was only possible with voluntary muscles. Now that this distinction has had to be dropped, the question has been asked if there is still any difference between classical and operant conditioning. It is in any case clear that "learning" is a fundamental biological process and, inasmuch as it is, the results of these experiments impinge on the distinction between "body" and "soul", "nature" and "nurture", and "instinct" and "culture".

CHANGE OF OBJECT

The alleged differences between classical and operant conditioning have also been put in doubt by other discoveries. So far I have been using the expression: "The rat presses the lever." This is not completely correct. Sometimes the rat does not press but chews the lever and lowers it as a result. In publications on the learning of rats this strange fact is often ignored. Yet we know that chewing the lever is a very common phenomenon, not so much from the publications as from advertisements for the Skinner box, for example: "The lever itself is made of stainless steel. It is 2 inches wide by $\frac{1}{2}$ inch thick. The end is rounded and smoothly finished. This, in addition to the thickness and hardness of the lever, prevents the rat from chewing and biting it. . . ." Thus important scientific data have often to be plucked from advertisements. Why is this an important fact? Before I come to it let me first mention something about other animals.

Breland and Breland worked with a racoon. This animal washes its food before it eats it. The two Brelands taught the racoon to drop a coin in a tray, whereupon it was rewarded with food. The cage contained a tray with water so that the racoon could wash the food it had been given. After some time, however, the racoon began to wash the coin instead. The coin, as it were, had been transformed into food. Similar phenomena have been discovered in the training of pigs and pigeons. Thus Moore found that, when a light was on to show the pigeons that they could now peck at the key to obtain food, they would peck at the light instead. If they were rewarded with water instead of food they would try to lick the light *and* the key. Rats, too, may chew at the lever when they are rewarded with food, and the first rat I rewarded with water proceeded to lick the lever. But this only happened once. All the other rats pressed the lever when they were rewarded with water. Now I do not own good Skinner boxes, and this may explain why I did not observe much chewing or licking of the lever.

The results of my experiments suggest that the change of object (coin "becomes" food) is not common among rats. Sevenster has claimed that this is precisely why the food-searching behaviour of the rat lends itself so excellently to conditioning: in the wild, the animal lives on a very variable supply of food. Perhaps this explains the great plasticity of the learning ability of these animals and also why the change of object in rats is not so pronounced as it is in pigeons (although it may turn out to be so if the experiments are performed with good Skinner boxes).

These experiments, too, have raised the question of whether or not there is a difference between classical and operant conditioning. When the dog hears the bell, it will begin to salivate once a conditioned reflex has been formed. The bell has, in a sense, been changed into food. What is the difference between the bell and the racoon's coin, which, in a sense, has also been converted into food?

SUPERSTITIOUS BEHAVIOUR

A strange phenomenon produced in the Skinner box is what Skinner himself has called superstitious behaviour. This type of behaviour appears chiefly when animals are given a reward without having done anything for it. Some rats start to chew their tails, others run around in circles, wash or groom themselves—in short, they obviously associate the reward with some action they performed accidentally immediately before receiving the reward, and not with the depression of the lever or the lighting up of a lamp. Skinner believes that similar reactions also occur among human beings: thus events which took place shortly before a disaster are often considered omens of that disaster. But then Skinner believes that all human behaviour can be understood from the results obtained with rats in the Skinner box. I shall discuss this highly controversial and dangerous approach in Chapter 9.

SOCIAL BEHAVIOUR IN THE SKINNER BOX

Very few people have inserted several rats into a single Skinner box, Mowrer (1940) being probably the first to do so. He first taught the rats to press the lever individually. Then he placed three rats in the box simultaneously. The animals pressed the lever in turn but were not properly rewarded every time, since one of the other animals would often reach the food tray first. During a second session with the same animals, Mowrer observed that there was a tendency to congregate round the food tray so that no food at all appeared. During subsequent sessions, one rat would continuously press the lever (Mowrer called this rat the "worker") and the two other rats would get the food (Mowrer called them the "parasites"). This situation persisted throughout all the subsequent sessions. The same rat remained the "worker" and only obtained its food when the other two animals were satiated. Hilary Box repeated these experiments in 1967. She, too, found that one rat developed into a worker and that the other two became parasites. She tried to predict which particular rat would develop into the worker. Did this depend on the extent of previous training? Was there a link with dominance? She was unable to discover any such links, partly because she failed to consider rat behaviour in general. For the rest she was very careful not to draw hasty conclusions. In particular, she considers it premature to speak of "co-operation" in a situation that has not yet been studied in any depth.

7.
THE CAGED RAT

An extraordinarily large variety of experiments has been conducted with the caged albino form of the brown rat, so much so that were I merely to list the names of these experiments I should have to cover scores of pages. Every day brings reports of new experiments. This very day I read a story in an evening paper about rats that were fed exclusively on apples and whose teeth had begun to rot quickly as a consequence. From this experiment, the journalist drew the altogether premature conclusion that eating apples is bad for the teeth of men. Hasty conclusions of this kind are drawn every day. Thus from the fact that rats turn to alcohol if they are made to inhale large quantities of carbon monoxide, one writer has concluded that people involved with road traffic who, after all, inhale a great many exhaust gases, also tend to turn into alcoholics. Now this may very well be the case, but it is quite impermissible to draw conclusions about human beings from experiments with rats—at best such experiments should encourage further studies.

A fairly full treatment of all the psychological experiments performed on rats would fill a book the size of the Bible. Here I can only refer the reader to Munn's *Handbook*, to the *Journal of comparative and physiological psychology* and to the *Journal of the experimental analysis of behaviour*. Gilbert and Sutherland's *Animal Discrimination Learning* gives an excellent survey of very interesting experiments into rats' powers of visual discrimination. In what follows I shall dwell on just two psychological experiments with rats that provide particularly striking illustrations of the approach used by animal psychologists and many other non-biologists, and that have both caused something of a stir.

THE ENRICHED ENVIRONMENT

Since the 1950s, Bennett, Diamond, Krech and Rosenzweig, followed by others, have performed a number of unusually interesting experiments. In particular, they have worked with an enriched environment, i.e. a substantial cage fitted with all sorts of toys such as a roundabout, a ladder, a whip, a swing, etc. Into the cage, they inserted ten to twelve young rats and kept them there from the 25th to the 105th day of their lives. A similar number of animals was placed into empty cages; here, each rat was kept in solitary confinement, in quiet surroundings in which other animals could neither be seen nor smelt. This was the impoverished environment.

On the hundred-and-fifth day of their lives all the animals were killed and their brains analysed. The brains of the animals from the enriched environmment proved to be different from the brains of the animals from the impoverished environment. The cerebrum of the former turned out to be more voluminous and heavier; the capillaries were wider, and there was greater enzyme (acetyl-cholinesterase and cholinesterase) activity. There were more glia cells; there was less DNA per milligram of tissue but the same amount of RNA so that the ratio of DNA to RNA was changed in favour of the RNA. All these changes pointed to an increased metabolic activity in the brain.

Later Globus, followed by Volkmar and Greenough, discovered that the number of dendrites (branches of nerve cells) per nerve cell was greater in animals from the enriched environment. Mollgaard found that the synapses of rats from the enriched environment were on average 50 per cent larger in diameter than synapses from the brains of animals that had grown up in empty cages.

The most important conclusion to be drawn from all these experiments is, of course, that the brain is moulded by what occurs in the external world. This plasticity is greatest while the animal is growing up but even when the brain is fully developed, changes can and do appear in the wake of further experience. In this field, particularly striking discoveries were also made with cats.

What precisely is the effect of changes in the brain on the behaviour and learning ability of rats? Students have compared

the learning ability of rats from an enriched environment with
the learning ability of rats from an impoverished environment,
but this work has not yielded spectacular results, possibly
because all the rats were tested under impoverished conditions.
I myself have raised rats in both types of environment, and have
observed the behaviour of the two groups. Young animals from
the enriched environment displayed a greater amount of exploratory and aggressive play behaviour in an observation cage.
During aggressive play behaviour, I sometimes noticed the
offensive upright posture and the beginnings of the sideways
posture. Unfortunately I was unable to investigate either at any
great length: in an enriched invironment, it should, in any case,
be possible to perform very wide-ranging experiments, not least
bearing on the general behaviour of rats.

BRAIN TRANSFERS AND TRANSPLANTS

If a flatworm is divided into two, a section with a head and a
section with a tail, the first section will grow a new tail and the
second section will grow a new head. Anyone who teaches the
undivided flatworm a new trick will notice that the two new
flatworms which emerge after the division will both be able to
perform that trick. This fact persuaded James V. McConnell and
collaborators to feed trained flatworms to hungry conspecifics
and to investigate whether the ingestion of the trained worm
enhanced the learning rate of the cannibal. This seemed to be
the case. Immediately afterwards the experiments were repeated
by others who used different controls and failed to obtain the
same results. The upshot has been a series of rather unfortunate
controversies. One can read it all in Chapter 3 of Ungar's
Molecular Mechanisms in Memory and Learning. In that
chapter, McConnell and Shelby discuss their own experiments
with flatworms, those of others, and the opinions of the various
critics. They believe firmly that there is transmission of learned
behaviour in the way they have described, a view that others
challenge just as firmly.

Needless to say, this famous experiment has been repeated
with animals other than flatworms. And naturally most workers
have used the only alternative with which they were familiar,

namely rats. The procedure is simple. A rat is taught a trick, killed and its brain (after appropriate treatment) injected into a rat that has not yet learned the trick. An attempt is then made to establish whether this rat can be taught the trick more quickly than, say, a rat injected with the brain of a conspecific that had not learned the trick but had shared the cage of a rat that had learned it. Such controls are absolutely indispensable.

This type of experiment was performed in countless learning situations. Oddly enough, it turned out that whereas one laboratory achieved wonderful results others drew a complete blank. Now this could have been easily explained had the experiments been unusually complicated and had they required a great deal of skill. However, this was by no means the case. Only the preparation of the brain extract demanded much skill and expensive apparatus but both could be repeated by any competent laboratory assistant.

With the help of Dr O. Wolthuis, a Dutch expert in brain transplants, I performed a similar experiment. It too, was described in the literature as having produced marvellous results. A cage with a grid floor contained a small platform away from the wall. On this platform I placed a rat, which jumped down after a few seconds. As soon as it did so, it received an electric shock from the grid floor. The rat learned very quickly to jump back on the platform. Once a rat has experienced such shocks it will not jump off the platform again in a hurry. To make doubly sure, I placed my rats in the cage twice a day and pushed them off the platform with the help of a plunger. During the process, the rat tried as hard as it could to cling to the platform or to the plunger. It lost every time. It fell on to the grid floor and then quickly jumped back on to the platform. During this experiment I acquired a heartfelt aversion to all such cruel experiments.

After the rats had been trained in this way for a week, they were decapitated and their brains injected into mice. There was, of course, a control group that was also placed in the experimental cage and pushed off from the platform but not given an electric shock. The brains of these rats were also injected into mice. Why this transfer from rat to mouse? Because the literature is full of good reports of such transfers. I tested the mice for their platform behaviour. I could not, of course, tell which mice had been injected with the brains of trained and which with the brains of untrained rats. The results were negative. The mice which had been injected with the brains of trained rats showed the same platform behaviour as did the other mice. No transfer of information could thus be detected. One might, moreover, ask oneself what precisely would have been transferred if an appreciable difference had been detected? Whoever uses such methods for a whole week to teach a rat to remain on a platform is committing something that looks surprisingly like ill-treatment. In this process chemical substances may be liberated which may produce muscular paralysis even in the mice that have this substance injected into them. In my case, there was no effect; but the literature is full of amazing results. Let me ask again: what, if anything, is being transferred?

I see no point in describing further transfer experiments. Often they are variations on a single theme. Ungar, who has published a giddying amount of positive results claims that the substance involved has been isolated. He has even synthesized it and sent it to other workers.

While he obtained positive results with it, however, the others failed to do so. Now this is not at all unusual: a few people often obtain positive results that others cannot duplicate. Most of the failures are, however, not published, so that there are a small number of positive and a large number of negative results. Perhaps it is best to cover this whole episode, which has caused such a stir in the newspapers and even in some textbooks, with a mantle of silence.

ONCE MORE: THE RAT AS EXPERIMENTAL ANIMAL

As early as 1949, F. A. Beach protested at a meeting of the Division of Experimental Psychology of the American Psychological Association against the exclusive use of rats as experimental animals. In his address, which was later published in *The American Psychologist* he said that it is essential to work with all sorts of animals, and, moreover, not to confine the investigations to conditioning and learning. The *Journal of*

Comparative and Physiological Psychology, he claimed, had better be called the *Journal of Rat Learning*, but "there are many who would object to this procedure because they appear to believe that in studying the rat they are studying all or nearly all that is important in behaviour. At least I suspect that this is the case. How else can one explain the fact that Professor Tolman's book *Purposive Behaviour in Animals and Men* deals primarily with learning and is dedicated to the white rat 'where perhaps most of all, the final credit or discredit belongs'. And how else are we to interpret Professor Skinner's 457-page opus which is based exclusively upon the performance of rats in bar-pressing situations but is entitled simply *The Behaviour of Organisms*". Is the rat, Beach went on to ask, the "representative species" Skinner claims it is? Beach himself had his doubts. Moreover, much too little attention had been paid to "instinctive behaviour". In this connection he praised Lorenz's studies of instinctive behaviour and described them as a hopeful development. Beach was right to think so, for it was Lorenz who ushered in an entirely new approach to the study of animal behaviour.

R. Boice and R. B. Lockard, too, have protested against the use of white rats. Lockard wrote in an article entitled "The albino rat: a defensible choice or a bad habit?" that white rats have been so inbred that they no longer resemble their wild conspecifics. There are anatomical changes: the liver, the heart, the kidneys, and the brain of the white rat are smaller than those of the wild brown rat, while the pituitary gland, the thymus and the thyroid gland are larger. He proposed the alternative of using primitive insectivores (tree shrews and moles) because these are closer to man than rats, man allegedly having evolved from proto-insectivores.

These protests led to a mass of investigations into the differences between wild and laboratory rats, and these studies were represented as important contributions to the understanding of the domestication process. But the wild rat, as I said earlier, is also a domesticated animal, so that if there are any differences between wild and white rats they cannot be solely due to the domestication of the white rat. Worse still, experimental psychologists are not in the habit of visiting refuse tips or river banks to study the normal behaviour of brown rats. The wild rats are instead brought to their laboratories, where these animals behave in a highly abnormal manner.

Nor can I agree with Lockard's contention that we should eschew work on wild rats which are so inbred that the results cannot be applied to man. For are human beings not inbred as well? And is the degree of inbreeding a proper standard of comparison, in any case? No, the whole principle of comparing human beings with rats is incorrect. And were we to work with moles and shrews instead, on the grounds that they are allegedly closer to man, we should merely come up with another series of ridiculous extrapolations, as witness much of the current comparative work on rhesus monkeys. These animals are usually kept in empty metal cages. If you observe them listlessly sitting about and read the reports of the nonsensical experiments to which they are being subjected, you will be hard put not to weep for them.

Of course I agree with Beach's claim that behavioural studies must not be confined to just one species of animal. As long, however, as experimental psychologists do not revise their methods and refuse to dispense with their grid floors, shuttle boxes, open fields, climbing poles and all the other instruments of torture, they are best left with their white rats. These animals are, in any case, better adapted to laboratory conditions and to men who dislike their experimental subjects than are rhesus monkeys or shrews.

Nor is that all. At a congress, I heard a lecture about the aggressive behaviour of rats under the influence of various drugs. The learned speaker discussed the merits of these drugs in some detail. But he spoke not a word about his method of recording the ensuing acts of aggression. During the discussion that followed I asked him about it. It appeared that he had placed a tape recorder near the cages of the experimental animals and that his recorder was switched on atuomatically every evening to register the screams of the poor beasts. A laboratory assistant counted the screams and from them gauged the intensity of the aggression. In fact, no one actually *saw* any aggressive behaviour. Now since many of the screams normally emitted during fights were suppressed by some of the drugs, it is quite possible that the animals kept fighting without the tape recorder registering anything. This might, of course, help to market the drugs in question as aggression suppressors!

In short, there is a tendency to dispense with direct observations, and, indeed, Skinner has referred slightingly to direct observations of experimental animals. Many experimental psychologists seem to look upon the observation of behaviour as a form of hard labour. They justify their aversion to direct observation by claiming that the human observer may make many mistakes and that he may also influence the behaviour of the experimental animals by his presence. These are valid objections, but they can be met, for instance, by the introduction of one-way mirrors through which several observers can watch the animals simultaneously and arrive at fairly objective assessments and descriptions of the behaviour patterns they have observed. The alternative is a completely automatic recording device. With such devices, however, a rat may die, fall on the registration button and thus ensure that the counters keep turning. In any case, many experimental psychologists rarely come eye to eye with their experimental animals. Laboratory assistants or stable hands place the animals into closed chambers in which the so-called experiments are conducted. Expensive apparatus collects the data, these are elaborated by computers, and the research worker merely looks at the mathematical results. To persuade the animals to perform tricks in their uncongenial chambers, the latter are of course furnished with the indispensable grid floors which, in turn, inspire a host of publications on, say, "shock-induced aggression". It strikes me that this sort of study can be compared with an investigation of the motives of china after the crate has been toppled over. One may then write a learned treatise on the behaviour of the shards without grasping for a single moment that shards may bring luck but never understanding.

8.
RAT CONTROL

AN OLD LEGEND

One evening in August 1953 two men performed a strange experiment on a deserted refuse tip in Altwarmbüchener Heide, Germany. One, an Englishman from Corby, famous throughout Germany for his radio talks, was there at the special invitation of a manufacturer of rat-traps. He was carrying a strangely shaped wooden tube. The other man, a well-known German rat expert, had chosen the site. Heywood, the Englishman, blew into his wooden tube, producing a strange sound . . . but we had best let Steiniger, the German, tell the story in his own words.

> The tone resembled certain sounds that occur in the repertoire of rats. It was, however, considerably louder, so that rats could hear it from far away. Heywood called it the "mating call" and intimated that it was different for males and females. As I wrote earlier, rats do not often emit sounds in connection with mating. If, after mating, the female bites the head of the male symbolically or actually, a kind of whistling can usually be detected. So far no one has been able to ascertain whether this sound is uttered by the biting female or by the male under attack. The sound resembles the tone emitted by Heywood's tube, except that the latter is louder. But it also resembles the terrified squeal of rats which are being threatened by others outside the mating context. In general one can say that Heywood's tone is not very specific in the sound repertoire of rats. Heywood blows every two to three minutes for an eighth to a third of a second, sometimes twice in succession. The resulting tone may be said to resemble that of a female mating with a number of males.
>
> During Heywood's experiments rats would emerge from their holes some five to ten minutes after the beginning of the experiment. However, it was too dark to study their behaviour precisely.

So much for Steiniger. Next morning at 5.30 the experiment was repeated. Once again a few rats appeared, but they did not run in the direction of the tube. For all that, "the rats were very attentive and seemed to be looking for something special," Steiniger added.

An attempt was being made here to repeat the exploits of the Pied Piper of Hamelin, with enough success to let Heywood share in some of the fame and magic powers of the rat-catcher. In fact, rat-catchers have often been the subjects of interesting books, and in England especially many of them have published their own memoirs. But I am not trying to write a comprehensive history of rat-catching (on which subject many weighty tomes could be compiled) and shall confine myself to a few brief remarks on the subject.

Man has fought rats ever since he first recorded their presence in the Middle Ages and perhaps even before then, though there is no means of telling. Over the centuries, even the reasons for fighting rats have remained largely unchanged. True, we no longer fight rats because we associate them with the devil or with witches, but now as then we know that they eat our crops, that their bite is dangerous, and that, when present in large numbers, they are associated with the spread of bubonic plague (although the role of the rat flea was not suspected until the nineteenth century). Rats must be exterminated, people have always argued, because they are dangerous: they transmit diseases and they bite.

THE DAMAGE CAUSED BY RATS

Attacks on human beings

In Table 12 I have listed a number of press reports published during the past two years of attacks by rats on human beings. Possibly these are all reports published in the "silly season", but a certain consistency—all the victims were children, drunkards and old people—suggests that some reliance may be placed on them.

To two of the reports I would like to add a marginal note. The Savona episode concerned a girl from a poor family. They lived in a basement and the mother would usually hunt rats in

Table 12. Press reports of children, drunkards, elderly people and others being bitten by rats

Date	Source	Place	Person bitten	Where bitten	Effects
May 1960	Nieuwe Noord-hollandsche Courant	Monnick-endam	child (5 yrs)	not specified	under treatment
Sept. 1960	De Tijd	Turin	man (74 yrs)	not specified	died
Sept. 1960	De Tijd	Savona	girl	half of nose bitten off	under treatment
1960	De Volkskrant	St John's, Canada	baby	head	died
1960	Rat en Muis	Waspik	elderly woman	face	not serious
23.3.1961	Rat en Muis	Drunen	girl	ankle	rat-bite fever
1962	Rat en Muis	Vollen-hoven	woman	nose	not serious
16.8.1962	Nieuwe Noord-hollandsche Courant	Brussels	two children	not specified	numerous wounds
28.2.1964	Nieuwe Noord-hollandsche Courant	Hoogkerk	man	three times in calf	not serious
1964	Rat en Muis	Waubach	child (3 yrs)	face	not specified
Nov. 1965	Rat en Muis	Roermond	child (8 months)	arm and shoulders	hospitalized
9.2.1967	De Tijd	Messina	student (21 yrs) after sleeping tablets	wrist and throat	died
May 1967	Rat en Muis	?	man, drunk (67 yrs)	nose	not specified
10.11.1967	De Telegraaf	Beesd	child (6 yrs)	arms and ears	under treatment
3.12.1969	Het Vrije Volk	Utrecht	two children	mouth and other parts	hospitalized
28.1.1971	Arnhemse Courant	Paris	baby	nose, chin and finger-tips gnawed off	died

the evening. On the night that the girl was bitten in the nose, the mother had been too tired to chase the rats away. An even sadder tale comes from Brussels. Here two children had been put to bed and tied down in an unsupervised hospital ward. They were quite helpless and they were attacked by rats. In most cases, the rats did not so much bite as chew. They were looking for food, and it was generally people who created the conditions in which the animals could chew undisturbed. It is precisely the poor who are their chief victims. Hence the special law passed by Lyndon B. Johnson—a law of which some silly Republicans tried to make fun.

Rats are undoubtedly harmful. The damage they cause to the economy is inestimable. In India, where there is hardly any rat control worth mentioning, the estimated rat population of $2\frac{1}{2}$ thousand million is said to consume 50 per cent of all crops; in the United States, where the number of rats is estimated as one for every two inhabitants, rats are said to consume 3 per cent of the crops. Between these two extremes lie various other countries: in the Netherlands I estimate the losses at about 5 per cent per annum. Taken over the whole world the loss is probably in the region of 10 per cent. But it should not be forgotten that these are mere estimates. In reality the figures are very likely much higher, if only because contaminated supplies must generally be destroyed. By way of illustration, let us look at some of the reports published by the journal *Rat en Muis*, which does not, of course, report everything that happens. During the last ten years it reported that over seven thousand chickens and two hundred pet birds were killed by rats, and that seven piglets were consumed wholly or in part. In Bingelrade, six young whippets were killed, and three chinchillas in Overdinkel. Another report mentions the killing of nineteen breeding rabbits, and a calf whose eye was gnawed out. There are also reports of damaged conveyor-belts, roofs, equipment, and so on, on chicken farms. Every year rats drag off countless eggs, sometimes immediately after they have been laid. Rats damage sewers, they chew through lead pipes; they have a special predilection for electricity and telephone cables. All this, once again, only gives an overall impression of the possible damage. The actual damage cannot be put into figures, the less so as rats seem to have an irrepressible urge to chew up almost any-thing—some even attacked the feet of elephants in Hagen-beck Zoo, following which the elephants had to be destroyed.

The transmission of disease

Rats can transmit diseases. This is certainly the most serious threat they pose. It is also something that people usually misunderstand. Nowadays, few of us bother about bubonic plague, but the figures show that the disease is spreading again, and in Vietnam there is talk of an epidemic. It is by no means inconceivable that the plague has "roused up its rats again and sent them forth", as Camus put it. Moreover, if an infected rat travels by plane, say from Vietnam to the United States, the plague can be spread very quickly to and by local rats.

Rats are supposed to play a part in the transmission of the following diseases: bubonic plague, Weil's disease, trichinosis, paratyphus, rat-bite fever, Haverhill fever, mud fever, foot-and-mouth disease, swine-fever, rabies, Newcastle disease, Aujeszky's disease (pseudo-rabies), toxoplasmosis, splenic fever and Bang's disease. Rats are well known to be agents in the dissemination of the first six diseases; in the remainder their involvement seems fairly certain. A short description of the principal diseases transmitted by rats follows.

Bubonic plague. Everyone who has read Camus's *La Peste* (The Plague) knows about the course of this disease. The glands in the groin, the armpits and throat swell into so-called buboes, become infected and eventually suppurate. There is a high fever, the eyes are red-rimmed and the patient has severe headaches. If there are a very large number of bacteria in the body the lungs may become infected as well, the result being pneumonic plague.

The plague bacillus is transmitted by the rat flea (*Xenopsylla cheopsis*), and possibly by other species of flea too. As soon as a rat dies of the plague, the flea leaves its body and looks for a new host, generally another rat. Occasionally the flea will jump on to a passing human being, bite him and thus introduce the plague bacillus into his blood. The resulting disease is not contagious, and can only be transmitted by flea-bites. As said earlier, bubonic plague can, however, turn into secondary pneumonic plague, which is readily spread by coughing. People who catch the germ contract primary pneumonic plague from which they can die within two to three days.

Bubonic plague has not disappeared, though epidemics of the kind known in the fourteenth to seventeenth centuries no ~rope) longer occur. The reason is not known[.] People have speculated ¹ᴾᵘᵇˡⁱᶜ ʰʸᵍⁱᵉⁿᵉ on the subject, some arguing that the brown rat has ousted the black rat, which transmits the disease. This seems improbable. All sorts of rodents can transmit the plague by means of their fleas, and without doubt the brown rat is one of them. Even squirrels can transmit the plague. The most bizarre theory on the disappearance of the plague is that put forward by J. H. van den Berg. He claims that the rise and decline of the plague are linked to the anatomical study of the human body begun in the fourteenth century. The first complete description of the anatomy of our body is said to have been given in about 1667, the date the plague disappeared from Europe (*Het menselijk lichaam*, 1; "The human body", 1.) It is, of course, impossible to prove or to disprove this very strange hypothesis.

Weil's Disease. Weil's disease is caused by *Leptospira ic-terohaemorrhagiae*. This bacillus dwells harmlessly in the blood of the rat and is excreted with the urine by its kidneys. The bacillus can live in water for a few minutes at 4°C. and up to a few hours at 20°C., and can enter the body of swimming animals or human beings through small wounds or through the mouth. After a fairly long incubation period, the victim sickens with cold shivers, a high temperature, jaundice, muscular pains, skin bleeding, red urine and acute inflammation of the kidneys. Not only swimmers, but sailors, fishermen, sewerage workers, reed cutters, and even firemen (through pumping contaminated water) can be infected. It is estimated that 30 per cent of our rats carry the germs, and since Weil's disease often has a fatal outcome, this percentage is most alarming. More-over, the widespread thermic pollution of rivers keeps the bacteria alive much longer than normal. For many years Leyden was notorious for its contaminated water. Not so long ago a student, delighted at having passed an examination, dived into the Rapenburg and died of Weil's disease.

Trichinosis. This disease is caused by a hairlike worm (trichina) no more than 1.5 to 3 mm. long. Its natural habitat is the bowel of the rat. If a pig catches and eats the rat, which happens quite often, the worm enters the bowel of the pig, where it bores itself into the intestinal wall and gives birth to young trichinae which travel from the bowel to the lymphatic vessels, eventually finding their way into the bloodstream and the muscular tissue. Here they coil up into spirals and encyst themselves, in which form they can remain alive for many years. If human beings eat uncooked pork (e.g. sausage meat) the trichinae may follow a similar path, as they do in the pig, though they generally prefer to settle in the diaphragm, the tongue, the larynx and the muscles of the eye. The patient is feverish, and suffers stomach aches, nausea, vomiting, diarrhoea, sweating and a red, swollen face.

Thanks to the scrupulous supervision of abbatoirs in Western Europe the danger of trichinosis is not nowadays very great. Moreover, heating pork to a temperature of 70°C. is sufficient to destroy the trichinae.

Paratyphoid. This disease is caused by *Salmonella enteritidis* var. *Danysz.* The symptoms are high fever, headaches and diarrhoea. The illness is usually quickly cured but is extremely unpleasant. The bacteria are transmitted by contaminated cattle and by preserved food processed in rat-infested plant (e.g. canned fish that is not cooked before or after canning). Duck and eels can also carry the bacteria.

Rat-bite fever. The name of this disease is misleading, for it can also be caused by the bite of monkeys, cats, parrots, weasels and squirrels. It results from infection with *Spirillum morsis muris.* The wound becomes inflamed, the lymph glands swell up and there is a skin eruption as with measles. The patient suffers from rapid variations in temperature, as large numbers of bacteria enter the bloodstream intermittently from the lymphatic vessels. These variations tend to exhaust the patient.

Haverhill fever. This disease also occurs after the bite of a rat and strongly resembles rat-bite fever, but it is caused by another bacillus, *Streptobacillus moniliformis*. The disease was named after a market town in East Anglia where an epidemic occurred in 1926.

Mud fever. This disease, which is caused by *Leptospirosis grippotyphosa*, slightly resembles Weil's disease. The bacillus is transmitted by field-mice, though rats are also mentioned as carriers. The disease is not very dangerous.

Newcastle disease. This contagious poultry disease resembling fowl plague is caused by a virus. Any moving object that has not been in the sun after infection with the virus can carry the disease. Since brown rats like to live in poultry farms and generally shun the sun they can play an important role in the transmission of the virus.

As very much less is known about the part played by rats in the other diseases I have mentioned—sometimes it is no more than suspicion—I shall not deal with them in detail. In all cases, it should be remembered that the danger is not so much the presence of rats as the presence of rats in large numbers. Very many other animals can transmit the same (and other) diseases, but because they are few in number they pose little if any threat to mankind.

The collapse of dykes

Because rats dig holes in dykes, the collapse of sea defences is one of the possible consequences of the presence of rats, although the musk-rat causes much more damage than does the brown rat. Writing in *Rat en Muis* in August 1967, H. de Vries has described the destruction by rats of a dyke in the Biesboch, Holland. It contained passages 3 feet long and even longer and it was perforated by them: the water had rushed through the holes and had carried the earth with it, thus hastening the collapse of the whole structure. The dyke was damaged over a total length of just under 500 feet, and the cost of repair in 1966 was about £100 per running yard. In addition there were several large holes in the body of the dyke each of which cost from £5,000 to £10,000 to repair.

METHODS OF CONTROL

History of rat control
Those who have read the above summary will realize that rats must be kept down. While the reasons for exterminating rats have changed little over the centuries, the methods of control have altered rapidly although one method—poisoning—has remained the favourite. Generally, special rat-catchers are employed for the purpose, though another favourite remedy used to be reading the Gospel of St John from three corners of the house, whereupon the rats were expected to disappear through the fourth, especially if it gave on to the house of a rich neighbour. People would also lay pieces of cloth sprinkled with holy water in the three corners of the house. In that case, too, the rats were expected to leave the house by the fourth corner.

Rats can also be driven off with the help of other rats. It is not an easy matter to catch a rat and to tie a bell around it, but at least it is not nearly as vicious as catching a rat and setting it alight, an episode we may read about in *I, Jan Cremer.* In that book, Jan Cremer tells us that all the rats on board a ship dived into the water after one of them had been doused in petrol and set ablaze. I doubt whether rats can be permanently frightened off in this way. What is certain, however, is that the squeals of a rat, whether it be burning or has been hung up by its tail, will make other rats extraordinarily nervous. An even more vicious variant of this method used to be employed in Eastern Europe: the anus of a captured rat would be sewn up and the animal then released. Its screams were said to be enough to persuade its fellows to disperse in a great hurry.

Attempts were also made to drive rats away with incantations. In a short book entitled *De muizen en ratten in de folklore* ("Mice and rats in folklore"), Jozef Cornelissen lists a number of them. One was in French and went as follows:
Quittez, quittez ces blés!
Allez, vous trouverez
dans le cave du curé
plus à boire qu'à manger.
There was also a widespread belief that rats could be chased away by noise, but though there is some indication that rats

may eschew the homes of brass-band players, this method, too. is not very effective. Nevertheless, it used to be customary in Friesland once or more times a year to gather together all the available pots and pans and other portable metal objects and to carry them in procession through the farmyard and house while banging them for all they were worth.

I have observed all too often myself how extremely sensitive rats are to noise. After the din of one day's drilling in a building in which the behaviour of rats was being studied, the work had to be abandoned for two or three days.

Attempts have also been made to infect rats with disease, a method that has produced such good results with rabbits. With rats, however, it was never a success. And what few reports there are about the administration of the pill to rats, for instance in *Nature*, 1969, 221, p. 906, give us little cause to hope that there has been any real breakthrough in this direction.

Fighting rats "biologically" has also failed. Thus when cats were specially imported into South Africa to control rats on large estates, it was found that the cats went after the few birds and left the rats alone. And man has eradicated almost all the rat's other natural enemies. This probably explains the several attempts to set rat against rat. One method tried was to shut up ten male rats together in a cage with no food. The weakest of the ten was eventually killed and consumed by the other nine rats. The process continued until one rat only was left. This highly aggressive male was sterilized and released, in the hope that it would kill all the males it encountered and that it would couple with all the females. (If a female is added to the group of starving animals, it is she that will survive, pregnant, of course.) The idea, however, was unsuccessful: the aggressive male was quickly killed by some of its stronger conspecifics outside.

Attempts have also been made to involve an entire population in the drive to exterminate rats. In Java, for instance, all those applying for a marriage licence were made to supply twenty-five rats' tails. The manufacture of artificial rats' tails (which were almost impossible to distinguish from the genuine article) blossomed into a flourishing industry. The authorities then asked for twenty-five dead rats, whereupon the Javanese began to breed them. In the Argentinian city of

Teresopolis (*Tijd-Maasbode*, 27 April 1965) a free cinema seat was offered for every five rats handed in. There were also periods in which Jews were forced to hand in rats' tails.

There is, however, only one method that really works, and that is poison.

Poisoning

Although rat extermination by poisoning is the most effective method of all, no single poison has ever done the job properly. Even the coumarin derivates in use at present, which for a long time were found most acceptable, are less effective than used to be believed, because rats can develop a resistance to them.

There are a number of reasons why poisoning rats is not a wholly satisfactory method of control. In the first place, people prefer to administer quick-acting poisons, so that they can actually see the results, before the rats have time to retire into their holes and to die there unseen. Now, quick-acting poisons are counter-productive, since they tend to nauseate the rats even before they have swallowed the fatal dose. As soon as an animal feels unwell, it stops eating, and defaecates or urinates over the bait, thus putting the rest of the pack on their guard. In this connection it should be pointed out that rat packs may include special "tasters", animals that test unknown food and also familiar food found in unfamiliar places. These rats are said to mark all food that does not agree with them, so that other rats will leave it untouched. There need be no special division of labour, however—a pack always includes young and curious animals; these youngsters automatically make up the "tasters". In any case, it is advisable to offer rats unpoisoned bait first so as to accustom them to it, and it must always be offered in the same place. Only then should the poison be mixed in.

In the second place there is one grave objection to the use of poisons: other animals and even human beings may take it by accident. People are usually warned off by the blue dye in the bait, but Chitty reports that, in England, one person a year is poisoned after eating poisoned rat food. Poisoned food can be placed in a so-called rat box to prevent dogs, cats and chickens from eating it. In Britain, this used to take the form of a small house with large eaves, under which the rats could sit safely while becoming accustomed to the house. Eventually they would make themselves very much at home inside. Thus Elton and Ranson report that in one such house, the so-called P_3 (Protected Poison Point), rats had brought in the following: a government form dealing with pigs, an empty snail shell, a piece of electric cable, a piece of tripe, the head of a starling, a piece of fish, an empty cigarette packet, one kilogram of cattle cake, and a number of other curious objects. The list will give the reader some idea of what strange objects rats will collect, a habit that may have some bearing on the origin of rat kings: rats may easily introduce sticky substances into their nests.

The following are some of the commonest rat poisons.

1. *Arsenious oxide* (As_2O_3) also known as "white arsenic". This is what is normally meant when people speak of rat poison, although apart from its legitimate use since the fourteenth century as a rat poison, white arsenic has been used since time immemorial by murderers. Napoleon is said to have swallowed a small quantity every day so as to become immune to the poison. The LD 50 (that is, the lethal dose which when administered to a group of animals, in this case rats, kills 50 per cent of the group) of As_2O_3 is 88 mg/kg of rat. The rats die slowly in their holes. As_2O_3 is poisonous to human beings, domestic pets and birds.

2. *Red squill.* This is a powder or extract made from the bulbs of the sea-onion (*Urginea maritima*), first mentioned by the ancient Greeks, but for its beneficial medical properties. It also has a long history as a rat poison, Chomel mentioning it as early as 1718. Since 1895 it has been a popular alternative to white arsenic. The sea-onion grows in most countries around the Mediterranean. The toxicity of the powder depends on the origin, colour and age of the bulb, on the time of harvesting and on the method of preparation. Red bulbs contain much more toxic material than white bulbs. The powder is most poisonous if the plants are harvested in autumn or winter. The LD 50 is 357 to 857 mg/kg for male rats; females are twice as sensitive to the poison. The toxic constituent is a glycoside (sciliroside). It is poisonous to man (although not to poultry), but since it causes vomiting, humans do not generally swallow it in fatal doses.

3. *Zinc phosphide* (Zn_3P_2). This is a black powder. The poisoned rats die quickly, often within an hour. The corpses lie on their stomachs, legs and tails extended. The LD 50 is 41.3 mg/kg. The powder is poisonous to man and to domestic pets.

4. *Barium carbonate* ($BaCO_3$). This substance has long been very popular as a rat poison, particularly in Germany, where it has been used since 1861. However, it has to be administered in very large quantities—the LD 50 is 693 mg/kg. Barium carbonate is poisonous to man, domestic animals and to poultry.

5. *Phosphorus* (P). Yellow phosphorus was first used as a rat poison in Germany during the last century. It is an unpractical poison because it poses a fire risk unless the bait is specially prepared. It is otherwise highly effective.

6. *Thallium sulphate* (Tl_2SO_4). This poison, first used in Germany in 1920, is tasteless and hence readily consumed by rats. It is, however, very expensive and may also be taken by other animals. It works very slowly. The LD 50 is approximately 20 mg/kg.

7. *Alpha-naphthylthiourea* ($C_{10}H_7NHCSNH_2$). This poison, usually known as ANTU, is an American post-war product. It is a greyish-white fine powder with a bitter taste. It works slowly and causes a form of pulmonary oedema, the rat, as it were, drowning on dry land. It is not effective against black rats. It is poisonous to man and domestic animals. The LD 50 is 6.9 mg/kg.

8. *Sodium fluoracetate* (FCH_2COONa). This is a white substance, usually called 1080. It has a characteristic smell and, like ANTU, is an American post-war product. The rat dies within from 45 to 240 minutes of taking the bait, the poison paralysing the respiratory organs. The LD 50 is approximately 2.5 mg/kg. Sodium fluoracetate is also very poisonous to other animals and to man.

9. *Strychnine.* This poison is not used very often. It is extremely toxic to human beings and domestic animals, and in most countries it cannot be sold over the counter. It has been used as a rat poison since 1640 and is extracted from the seeds of *Strychnos nux-vomica*, and other trees of the same genus. It works very quickly, paralysing the central nervous system.

Because of the bitter taste it is not readily taken up by rats. The LD 50 is 4.8 mg/kg.

10. *Coumarin powder* (Warfarin or Warf 42), or 3-alpha-acetonyl-benzyl-4-hydroxycoumarin. This substance is an anti-coagulant and works by lowering the pro-thrombin level of the blood. In addition, the capillaries are rendered porous so that the poisoned rat dies of internal bleeding. It dies painlessly, not realizing that it has been poisoned, not least because it dies so slowly. The substance works cumulatively and is dangerous to all warm-blooded animals, including pigs in particular. If animals or human beings are poisoned with Warfarin, the effect can be neutralized with Vitamin K.

Even rats that have been affected by coumarin continue to take the bait, proving just how effective this substance is—it produces no feeling of nausea, unlike so many other poisons rats quickly learn to avoid. Thus if rats are offered ordinary food and subsequently injected with an emetic, they will not touch again any of the food they ate before the emetic was administered, obviously associating their nauseous feeling with the food that happens to be in their stomach at the time. None of this applies to coumarin.

Coumarin bait is usually made from a mixture of oats and vegetable oil to which 0.025 per cent coumarin has been added. Large quantities are best prepared in a small concrete-mixer.

Coumarin and other anti-coagulant substances such as diphacinone, chlorophacinone, fumarin and coumatralyl are all very effective rat poisons. However, in 1958 Boyle discovered an area in Scotland between Edinburgh and Glasgow in which rats were or had become resistant to coumarin. In 1960 a similar area was discovered in Wales, and in subsequent years various small areas were discovered in which rats were resistant to coumarin. The Warfarin resistance of the rats from Wales has been investigated most thoroughly of all. It is not entirely certain whether resistance results from unusually fast excretion of the poison, or whether the Vitamin K in the blood of resistant rats continues to act despite the presence of coumarin. Drummond believes that the second explanation is the correct one. In any case, resistant rats cannot be poisoned with a hundred times the quantity of coumarin that suffices to kill normal rats. From a study by Greaves and Ayres it would appear further that the descendants of resistant rats crossed with non-resistant rats are also resistant and that 75 per cent of the third generation are resistant. In other words resistance is genetically determined and, moreover, dominant over non-resistance. This means that the number of rats which are resistant must quickly increase. Now, on the basis of the spread of coumarin resistance, it has been established that rats advance at a rate of twenty miles every six years. It has also been shown that, in many cases, coumarin resistance also means immunity from other anti-coagulants. Hence it is quite conceivable that, in a number of years, the entire rat population will be resistant to coumarin. It will certainly be some time before this comes to pass, but because coumarin has become the preferred rat poison, we are, in fact, breeding a resistant population selectively.

Many other rat poisons are also in use, including particularly castrix, fluoroacetamide, toxaphene, endrin, norbormide, alphacholoralose and gophacide. There are also regular reports of such new poisons as Ratcate MCN-1025. The names alone will suffice here; a detailed treatment would try the reader's patience too far.

PREVENTING RAT INFESTATION

Much more important than the eradication of rats is taking of preventive steps against rat infestations. Rats will congregate as soon as a suitable biotope is created. Poisons may kill some of them, but other rats will quickly come to take their place if nothing is done to stop them. Now, it is extremely difficult to keep rats out altogether, although many methods go a long way towards achieving that end. Thus a lavatory can be so plumbed that sewer rats are excluded by a water trap. Similarly, drain-pipes can be closed off with a grate. Ivy can be cut away from window frames, ventilation holes can be made very small and cavity walls so constructed that rats cannot climb up inside them. Unfortunately, however, society tends increasingly to create ideal habitats for rats. New refuse tips are set up, large new towns are built with those necessities for modern living, ponds and open green spaces, set around tower blocks with cellars that no one bothers to look at and that are littered with all sorts of rubbish. The ponds are stocked with ducks that are fed almost incessantly so that a great deal of food is left over for the rats. Moreover, rats find conditions on the so-called factory farms that have sprung up everywhere well suited to them. It sometimes looks as if we are going about things the wrong way. I shall be returning to this point.

RATS IN SEWERS

There are two sewerage systems. In the separate system, storm water is carried separately from sewage; in the combined system, storm water and sewage are carried together. Both systems have advantages and disadvantages. The combined system has the advantage that the drains are regularly flushed out with rainwater. However, the separate system is better if the system is joined to a sewage treatment plant, because it obviates the unnecessary purification of the storm water. As far as rat control is concerned, the combined system is preferable since, after cloudbursts, many rats will be drowned in the sewers.

The sewers to which lavatories are joined usually have a diameter of 8 to 10 inches. These run into larger pipes, though not large enough for a man to walk inside—for that a diameter of $3\frac{1}{4}$ feet or more is required and then the man will be bent double. It is essential for men to be able to negotiate the pipes because the system has to be regularly checked, and because solid deposits have to be removed. To that end, special shafts are sunk wherever pipes cross over or join, and also at intervals of about 200 feet in all the straight sections.

Sewers can teem with rats. In London, it is estimated that there are five hundred sewer rats to every running mile of sewer. The rats can come up through damaged drainpipes, through emergency exits, through damaged manhole covers or through breaks in the sewers themselves. Some will even brave water traps, so much so that, in Rotterdam, there are more than one hundred reports a year of rats having entered houses in that way.

Rats are not evenly distributed through the sewers, but

congregate where most of the food is to be found. Cracks in the sewers enable them to dig holes in the soil beyond; they also nest in the dead-ends and in the emergency exits, which are generally dry. Sewers are particularly well stocked with rats near abbatoirs, hotels and food factories and beneath blocks of flats.

It is important in the biological study of sewer rats to establish whether or not they live separately from the rest of the rat population; according to Bentley and Becker they do, but according to Barnett they do not. I tend to take the view that there are separate populations. The environment of the sewers differs too much from the environment outside for a single population to adapt itself successfully to both types. Nevertheless there will always·be sewer rats that can join surface packs and vice versa. Sewer rats live in an environment that is constantly dark and has a high humidity and an even temperature. In winter the temperature is usually higher than above ground (it never freezes) and in the summer it is lower. The high humidity explains why no black rats have ever been found in sewers.

The construction of good sewerage systems dates from the beginning of the nineteenth century. I should not be at all surprised if the construction of a biotope so excellently suited to the brown rat contributed to a large degree to the proliferation of this rodent—the unprecedented spread of the brown rat also occurred in the last century.

How well suited this biotope is to the brown rat may be gathered from the following figures. Inside the sewers, rats propagate their kind throughout the year. Barnett found that 27.3 per cent of females captured in sewers were pregnant, and Leslie found that 29.7 per cent of the females captured in winter were pregnant. In the same location, only 4.3 per cent of the females living above the ground were pregnant. Peters counted ten pregnant and three suckling females in a total of forty females, that is, 32.5 per cent. Telle found that throughout the year 25 per cent of the females were pregnant, while outside the sewers less than 5 per cent of the females were pregnant in the winter. In the summer more females above ground were pregnant (40 per cent) than females in the sewers (25 per cent), but over the whole year there were more pregnancies in the sewers than above ground in two groups of rats of the same size.

Admittedly, Telle found that the females in the sewers gave birth to smaller litters, namely 4.5 young per litter, than the females living above ground, who produced 6.33 young per litter. From this difference he deduced that conditions below ground were less favourable than conditions above ground. This strikes me as an unacceptable conclusion. Large litters are less favourable than small litters because, as I said earlier, the young in a small litter have a much better chance of survival. Above ground it is of course an advantage if, during the summer when there is a great deal of food about, the litters are large. But in the sewers, where conditions are fairly static, it is much better if small litters are born regularly, month after month. Moreover, a smaller number of young per litter may also indicate that there is a greater population density in the sewers, which would point to more favourable conditions than prevail above ground.

The eradication of sewer rats is more of an uphill fight than the eradication of rats elsewhere. In the United States young alligators are sometimes kept as domestic pets. As soon as the animals grow too large for comfort they are flushed down the lavatory. These alligators are now said to be present in US sewers in large numbers and to prey there on rats. According to my sources, it is no longer safe for men to enter these sewers. If the story is true, then a good method of exterminating sewer rats will have been invented, albeit by chance. But alligators are particularly sensitive to environmental pollution, so that it seems much more likely to me that most of the animals flushed away perish miserably.

In the sewers, too, rodent exterminators work with poisons. Above ground, the poison bait is usually placed near a rat run after all other food has been carefully removed. Sewers, by contrast, are so dark that it is impossible to locate runs, the less so as the constant supply of sewage, bringing fresh supplies of food, keeps altering the situation. Even if one should nevertheless succeed in poisoning some sewer rats, it is only a matter of a short time before the loss will have been made good with fresh litters. It is, after all, impossible to poison them all, and killing a few merely stimulates propagation among the rest. Thus Barnett reduced the rat population in a London sewer with sodium fluoracetate to 10 per cent of the original number.

After six months there were as many rats as before. In another sewer, 50 per cent of the animals died; after three months the original number had been doubled. No wonder that so many authors (including Bentley) take a pessimistic view.

Telle reports that sewer rats ignore or eat only small amounts of the poison bait. Becker states that after five days the poison in the damp sewers becomes covered with fungus, and no rat will eat mouldy food. To prevent mould, a solution of 0.1 per cent dehydro-acetic acid, of 0.25 per cent paranitrophenol, or of orthonitrophenol, is added to the bait, but this changes the taste of the bait and renders the rats highly suspicious. In addition, bacteria in the damp sewers render the food sour and unpalatable to rats. Again, since sudden rain storms wash the bait away, it is often placed in small containers and suspended above the water line from the steps in the shafts. The rats, however, ignore it. Little floating vessels have even been made to carry the bait, but the rats merely used them as Noah's Arks during rain storms. When the bait is mounted in blocks of paraffin wax (no mould, no bacteria) and the blocks suspended from the shaft steps, the rats also ignore it. Better results have been obtained with bait suspended in plastic bags, but this is not a very effective method. Attempts have been made to close off side branches and to drown all the rats inside, but those left in the main branches quickly make up the losses. Sewers have been pumped full of poisonous gases, although not without risk to the men manning the pumps and to householders using their lavatories at the time. When poisonous foam was used (Schürmeyer), it was washed away during rain storms and killed the fish in canals and rivers. There are, however, a few positive aspects of rat control in sewers. In sewers, after all, all kinds of poison can be used—there are no other animals at risk. It is also possible for the toxic waste products of certain modern industries suddenly to eradicate enormous quantities of rats underground. Nevertheless, rodent control remains a difficult task and is, moreover, very hard work. Over and over again, the officer must cover the sixty yards between manholes, lift up a heavy lid and climb down into the sewers to deposit the poison. It is not long before he runs straight on to the National Health—he is broken by the constant climbing up and down of the shafts. For this reason Chitty has constructed an ingenious device for depositing poison at depth. Simply put, it is a box with two cords; the first serves to lower the box, the second opens it in such a way that the poison bait falls out at a suitable place.

Although attempts to exterminate sewer rats are world-wide, Kuhn tells us that in Zurich, a small number (a probable euphemism) is deliberately spared to help surface workers detect breaks in the conduits (through which the rats can come up). In Zurich, 60,000 cubic metres of ground water is daily pumped up for the preparation of drinking water and it is therefore of very great importance that the ground water should not be polluted with seeping sewage. Kuhn reports that, with the help of the rats, at least two dozen breaks are detected annually.

THE JOURNAL "RAT EN MUIS"

In the Netherlands there is a unique journal devoted exclusively to the control of rats, mice and insects. It is entitled *Rat en Muis* ("Rat and Mouse") and is published by the Department of Vermin Control (which falls under the Ministry of National Health, Chief Inspectorate of Environmental Hygiene, in Wageningen). The journal is sent out to 4,600 individuals and authorities in the Netherlands concerned with rodent extermination. Apart from highly interesting information, the journal generally publishes magnificent photographs, of which a number have been used in this book. The fact that so excellent a journal is published in the first place shows how much importance the Dutch authorities attach to rodent extermination. Almost every local or district council in the Netherlands, as in other progressive countries, employs part-time or full time rodent exterminators who will do this work free of charge.

It has been most noticeable over the past few years that *Rat en Muis* has eschewed the blood-curdling propaganda that usually goes hand in hand with calls for rat control. This strikes me as an altogether praiseworthy development. We have to fight rats because there are too many of them, not because they are monsters, for they are nothing of the kind. I myself can only kill rats with a bad conscience; I feel that mankind blames rats for its own shortcomings, often to the point of blindness. Thus when Mijnheer W. Boogaard, an official pest control officer in Alphen on the Rijnm was asked why he had started to exterminate rats, he replied: "I have always been interested in animals."

NOT THE RAT BUT MAN IS TO BLAME

The control of rats puts one in mind of a never-ending game of chess. The rat is black and plays like Petrosian—it defends its position by constant frustration of the opponent's moves. Thus when white moves his pawns (poisons) black blocks white's major pieces (the methods of keeping rats at bay) or forces them onto unprotected squares. In the sewers, white has to play blind chess, which is even more difficult. And when white threatened mate with coumarin, black came up with a brilliant counter-move (coumarin-resistance). In addition black is careful to retain all its pawns (its cautious approach to new food, etc.) and uses its major pieces (its ability to go everywhere and to adapt itself to most conditions) very effectively. Moreover, black keeps making new queens, producing litter after litter of new opponents.

It is man himself who keeps supplying the chessboards. If he did not build sewers, or houses with cavity walls, have stagnant ponds with overgrown banks, docks with warehouses, lumber rooms, poultry farms, unused basements in blocks of flats, pigsties or abbatoirs, the brown rat would live as inconspicuous an existence as the squirrel. It is man who creates the biotope, who produces the conditions in which rats, starlings, mice, house flies and other pests can flourish and spread.

Some time ago I worked on a film in the docks of Amsterdam. Two rats were required for a feature film, and I had been asked to supply them. In the vast shed in which we were filming, large crates had been raised off the ground, ready for moving with forklift trucks. Now the spaces beneath the crates made ideal homes for rats: there was excellent protection since no one could get near them or even see them. No wonder the shed was teeming with rats. The producer did not need me; all he had to do was to wait patiently one evening, camera at the ready. That is one example; there are scores of others for those who care to look. The rat is assisted at every turn.

In short, the rat has man to thank for its success (its presence in large numbers throughout the world). It is ironic that man calls the rat harmful and noxious. Man himself is harmful: he destroys nature and thus creates favourable conditions for rats. Rats spread diseases because we ourselves create the conditions in which they can multiply and because we eradicate their natural enemies. Whoever despoils nature has to pay a price. It is only fit and proper that we should try to control rats because, in so doing, we rectify a mistake that we have made and that to a certain extent we continue to make. It is quite wrong, how-ever, to fight the rat with unfair argument. People should not be deliberately frightened with films of rats about to attack babies. Black propaganda invariably turns against its makers in the end.

Certainly, the rat is harmful. But there are other four-footed creatures that are much more harmful and yet not subject to vicious slanders. There is one in particular, for which man lovingly builds runs and shelters above and below ground, and for which he levels entire city blocks until they could be bomb-sites. The damage this four-footed monster does is many thousand times greater than the damage caused by the rat. How many people in your neighbourhood do you know who have died because of rats? I knew one student who died from Weil's disease after he had jumped into water infected by rats. But no less than five people in my closest circle have died as a result of meeting the other four-footer, and others have become lifelong invalids. In the Netherlands it is not the rat that has 3,000 dead and 20,000 seriously injured on its conscience every year. The rat does not lay waste to whole cities and the last vestiges of nature. And yet the rat is pursued as if it did all that, and the other is pampered like a pet. The effects of the motor-car can only be compared with those of the plague-carrying rat in the fourteenth, fifteenth, sixteenth and seventeenth centuries, except that the plague was merciful enough not to leave any horribly maimed survivors: one either dies of the plague or is cured. A campaign against rats is an absurdity while there is no campaign against the motor-car.

9.
MAN AND RAT

In his *Marius wil niet in Joegoslavië wonen* ("Marius does not want to live in Yugoslavia"), K. van het Reve tells of German, Russian and Canadian soldiers who ignored perfectly good water-closets in their billets and deposited their excrement, respectively, in a cellar, a bath and on the floor. Van het Reve gives no explanation of this strange behaviour. An ethologist, however, is put in mind of a hippopotamus, an animal that marks its territory with dung. Could it be that these soldiers used similar methods to lay claim to the houses in which they were quartered on enemy territory?

It is particularly tempting to explain just those forms of behaviour that cannot be easily expressed in the usual psychological terms, by examples from the animal world. Thus, why do we look round at our faeces as soon as we have made them? Why do we stand as far away from other people as we can in a confined space such as a lift (whenever someone leaves a lift, or someone else come in, positions are immediately re-shuffled so that everyone is again as far away from everybody else as possible)? Why do we feel uncomfortable if someone looks us straight in the eye? Why do we scratch our heads when we are embarrassed?

And why have there been so few attempts to answer this sort of question? Because we gather what information we can about human behaviour almost exclusively from conversations with and about others. It takes an original thinker, for instance F. J. J. Buytendijk (*inter alia* in his *Algemene theorie der menselijke houding en beweging*; General theory of human attitudes and movements) to delve more deeply into the problem.

Because the behaviour of animals, unlike that of human beings, cannot be studied by means of conversation, ethologists are forced to observe gestures and actions and, as a result, they have brought to light many a fact that may also help to explain the non-verbal behaviour of men. The study of animal behaviour can thus serve as a kind of aha! experience, as Gestaltists call it. In particular, certain obscure aspects of human behaviour suddenly make sense in the wake of observations of animals. Because of these aha! experiences, ethology has become extremely popular almost overnight, as witness the best-selling works of Morris, Lorenz, Eibl-Eibesfeldt and Wickler. Some ethologists (Morris, Lorenz) and some other writers, for instance the playwright-journalist Robert Ardrey and the biologist Dick Hillenius have, in their enthusiasm, put forward such sweeping generalizations about human behaviour that more cautious ethologists have felt compelled to take them to task. Thus Hinde and Tinbergen were among the first to question the idea that all human behaviour can be explained in terms of animal behaviour. But while Hillenius, to take but one of the new school, may have gone too far, I do not think that many of the critiques levelled against him are altogether fair. We should rather look upon Hillenius as an intrepid pioneer in the borderland of animal and human behaviour, as one whose ideas are not so much scientific dogmas as poetic inspirations. As for Ardrey, I have little time for his writings; they are full of misleading information about animal behaviour, and his conclusions are extravagant in the extreme.

Rats—and how could it be otherwise—play an important role in this wole confused story. Long before the first ethologist put his ideas about the behaviour of men and animals on paper, B. F. Skinner had already "explained" human behaviour in terms of experiments with rats in the cages called after him. Rats also appear in Konrad Lorenz's *On Aggression*, in Dick Hellenius's *Plaatselijke godjes*, in the two Russells' *Violence, Monkeys and Men*, in Ardrey's writings, and in all sorts of monographs on aggression and population. Are there indeed parallels in the behaviour of men and rats?

PARALLELS IN THE BEHAVIOUR OF MEN AND RATS

Are rats men in miniature? Whoever reads Skinner's extremely lucid expositions discovers that the writer does indeed consider human beings as complicated rats We have at any rate one thing (which Skinner does not mention) in common with rats: we exist in very large numbers. In the spring of 1973, the World Health Organization reported that there were then on earth as many rats as there were human beings, namely 3.7 thousand million. If one considers how difficult it is to determine the precise size of just one rat pack, then it is obvious that this number must be a shot in the dark. All we know is that there are millions upon millions of rats, but there might just as well be twice 3.7 thousand million.

We also have in common with rats the ability to adapt well to all sorts of circumstances, which means that we are not highly specialized. Neither rat nor man can run as fast as an antelope, see as far as a hawk, smell as keenly as a dog or swim and dive as well as an otter. The more specialized an animal the more limited its possibilities. Rats and men (just like sparrows, house mice, starlings, flies, and so on) are, as Lorenz put it, specialized in not being specialized.

Yet very little has been written on this almost incredible power of adaptation to changed conditions. Writers about the behaviour of men and animals generally dwell on two subjects: sexuality and aggression, as witness the title of one of Eibl-Eibesfeldt's books: *Liebe und Hass* ("Love and hatred"). True, Desmond Morris has covered a broader spectrum of human behaviour in his *The Naked Ape*, but his theories lose some of their importance because his model was a naked English gentleman. For the rest, in his *The Human Zoo* and *Intimate Behaviour* he, too, confines himself to the favourite topics of sexuality and aggression.

Why these two subjects time and again? Why nothing about feeding, evacuation, grooming, the daily rhythm, waking, playing, or the relationship of mother to child? As soon as one writes or speaks about aggression and sexuality one shoulders the top-heavy ideological mortgage that rests on both these subjects, whether one wants to or not. Now it is precisely because the subjects on which one writes and speaks are so heavily charged that one writes and speaks about them in the first place, mistaken though this approach is bound to be.

In rats, very little is known about aggression. Observations in the wild and in the laboratory contradict each other. There are, it is true, some indications about the function of aggression in these animals. Rats defend their territory and older animals drive off younger animals, especially males. As a result the species is encouraged to spread further afield. Aggression may also occur when male rats approach one another sexually. But it is not know whether aggressive rats can be bred selectively, as sticklebacks can. Thus while aggression is genetically determined in sticklebacks, we do not know what happens in rats. Nor do we know a great deal about the structure of aggressive behaviour in rats, and there are highly conflicting data on whether aggression increases or decreases after protracted isolation. Nothing is known about possible habituation to aggressive stimuli and what data we have on aggression and overpopulation in cages can be interpreted in very different ways. And what applies to rats probably applies to many other animals as well. In short, aggressive behaviour has still been so little investigated that writing on aggression must be pure speculation. This may explain why Lorenz's views differ so widely from those of the two Russells.

There are, however, some positive parallels in the behaviour of men and rats. The playful behaviour of children and young rats (and of young monkeys and young bears) has a comparable structure and also develops along comparable lines. Hunter and hunted change continuously; there are many collisions without injuries. There is also one characteristic difference. Young rats (and young monkeys) are handicapped in later life if they are not allowed to play in youth. One would expect children to bear similar psychological scars and this is, indeed, what Ardrey believes does happen. In fact, however, children who cannot play (for instance those who have no brothers or sisters) adapt themselves to the situation by developing, as Blurton-Jones says, into ''verbalists''. They are and remain talkers, their whole life long. This example shows how dangerous it is to extrapolate from the behaviour of animals, for instance from that of the hippopotamus mentioned at the beginning of this chapter; at most, the behaviour of the hippopotamus can be used as a working hypothesis whose validity has to be checked.

A second parallel in the behaviour of rats and men may perhaps be found in the enriched environment. Is intelligence a hereditary characteristic? From studies with rats it would appear that the environment in which the animals live greatly influences the structure and activity of the brain. The same might be true of human beings. Unfortunately, this hypothesis cannot be verified as easily in men as it can in rats, though I think it highly probable that comparable changes in structure and activity may appear in both species. In the impoverished environment represented by a town apartment, the brain no doubt obtains less stimulation than it does in an old-fashioned house with an attic and a garden.

A third parallel may perhaps be found in the learning of rats and men. Rats can learn to control their heartbeat, as can adepts of yoga. This similarity will have to be investigated in closer detail, but if men can, indeed, be encouraged to develop this capacity, new therapeutic horizons will have been opened up, especially for patients with a heart condition. All this lies, of course, far in the future.

If, moreover, one considers how much brown and black rats, though closely related, differ from each other, one grows all the more reluctant to draw parallels between rat and human behaviour in general, and aggressive behaviour in particular.

KONRAD LORENZ AND DICK HILLENIUS

By and large there are three theories about the origins of aggression.

1. Aggressive behaviour is genetically determined.

2. Aggressive behaviour is acquired.

3. Aggressive behaviour is a response to frustration (for instance, frustration caused by overpopulation).

The first theory is upheld by Lorenz and Hillenius; the second is upheld by Skinner and his disciples in the United States, and the third theory is upheld by the two Russells and Morris.

What little is known about the aggressive behaviour of rats would seem to corroborate Lorenz's theory. Rats that have been reared in complete isolation (and which could not, therefore, have learnt aggressive behaviour patterns by interaction with conspecifics) admittedly show less aggressive behaviour than other rats, but the structure of that behaviour is something I, for one, cannot distinguish from that of normal aggressive behaviour. Moreover white rats that have grown up normally and that have never shown true aggressive behaviour can, under the influence of certain drugs, display all the aggressive behaviour patterns characteristic of wild brown rats. It is often said that aggressive behaviour among white rats has been selected out (which, of course, implies that aggression has a genetic basis). This is completely untrue. White rats can fight like brown rats even without prior training. This, to my mind at least, suggests that aggressive behaviour patterns are genetically programmed.

Among normal rats—and that, too, agrees with Lorenz's view—aggression has a positive function (see p. 96). Moreover (and Lorenz writes about this at length) the behavioural repertoire of rats includes a signal that inhibits the opponent from launching further attacks, namely the emission of ultrasonic squeaks (see page 103).

Lorenz does not describe the behaviour of rats in neutral terms, however; on the contrary he mentions rats as an example of wild aggression. Yet nothing so far discovered in nature supports that view. What Lorenz and Hillenius do is to engage in a superior kind of gossip about rats, which, like gossip about human beings, is the result of hearsay information. One can only write about rats as Lorenz and Hillenius do if one has never performed experiments with these animals. And if one writes about experiments that one has never actually witnessed, one is bound to come up with all sorts of misconceptions and misinterpretations.

THE RAT CAGE IS A MONKEY HILL

Calhoun performed six experiments in his "quarter-acre enclosure". During each of the first three experiments he placed thirty-two rats inside the enclosure, which was partitioned into four compartments, two of which were joined by a kind of bridge. The compartments also contained towers that the rats could climb. In the second series of experiments Calhoun began with fifty-seven rats. He himself observed the very first group for twenty-seven months, at the end of which he was left with 150 adult rats. If the animals had mated normally there might have been as many as 5,000. Why were there only 150 animals? All sorts of mishaps occurred and even if the animals were born normally the mothers usually did not tend their young. Among the males Calhoun observed cannibalism and homosexuality. The food supplies near the centre of the cage were defended by a number of dominant males, who molested all the other animals. Young animals were pursued mercilessly; baby rats were killed. Haanstra has taken a film of similar behaviour in (so-called) densely populated mouse cages. C. and W. Russell quote Calhoun's experiment and observe that from it and similar observations it would appear that an animal society turns vicious whenever the population density becomes too great, and that this type of violence helps to keep the population down. This, they suggest, may well be the basis of violence among animals.

This explanation sounds convincing, but what are we to make of the following facts?

a) In nature more than 150 rats can live peacefully side by side in a territory very much smaller than a quarter of an acre.

b) In three of the six experiments described by Calhoun the situation was anything but a "behavioural sink", even though in these experiments, too, the animals had great difficulty in propagating and rearing their young.

c) If ten rats that have co-existed peacefully in a crowded cage measuring 24 by 16 inches are transferred to a space measuring 6 by $1\frac{1}{2}$ feet, they become and remain extraordinarily aggressive. As a result of this transfer—which I have carried out myself many times—the population density becomes smaller, not larger. Why then the aggression? After a few days all the animals bear fresh and old wounds. Much the same thing happens with chickens (Brantas) and field mice (van den Hoövel). The following explanation has been advanced by the same workers. In a small cage, animals cannot establish a territory, hence there is nothing to defend and no display of aggression. In a very large cage, by contrast, the animals can establish territories but these are so large that aggression is confined to harmless skirmishes in the border region. In a medium-sized cage, the territories they establish are not large enough, with the result that the animals try continuously to extend them—hence the bloody conflicts. I cannot say that this explanation satisfies me, for I have found that rats do not establish territories, however small, in a cage measuring 6 by $1\frac{1}{2}$ feet. I think it much more likely, therefore, that, in medium-sized cages; aggression is due to the animals having a fair chance of escaping but not of getting away completely. During one of my experiments I placed a small open cage inside the medium-sized cage. Two males began to fight and one made for the small cage. The other one followed. Inside, they faced one another, but their aggression quickly froze into the offensive upright posture. Then the animals flagged and soon afterwards they were both sleeping peacefully side by side. This is no more than an anecdote. But on the basis of this and similar observations I am inclined to believe that aggression does not occur if the animals can escape (large cages) or if they have no chance at all of escaping (small cages). Aggression only occurs if the animals can escape but not so far away that the other gives up the pursuit. Aggression is then not a consequence of overpopulation but of an interaction between being chased and not having enough room to escape. The behaviour of monkeys on an artificial hill probably has a similar explanation. Be that as it may: it is premature to conclude from this type of experiment that aggression is a consequence of over-population.

B. F. SKINNER

". . . He is dissatisfied or discouraged (*he is seldom reinforced, and as a result his behaviour undergoes extinction*); he is frustrated (*extinction is accompanied by emotional responses*); . . . there is nothing he wants to do or enjoys doing well, he has no feeling of craftsmanship, no sense of leading a purposeful life, no sense of accomplishment (*he is rarely reinforced for doing anything*); he feels guilty or ashamed (*he has previously been punished for idleness or failure* . . .); he is disappointed in himself or disgusted with himself (*he is no longer reinforced by the admiration of others and the extinction which follows has emotional effects*) . . ." This is a quotation from *Beyond*

Freedom and Dignity by B. F. Skinner. The author is describing a young man whose world has suddenly changed for the worse and whose responses can be expressed in terms of the operant behaviour of rats in Skinner cages (the phrases in brackets). This sort of prose arouses special emotional responses in me, not least the responses of abhorrence, dismay, revulsion and bewilderment. The quotation shows clearly to what extent Skinner feels free to describe the behaviour of human beings in terms of reward, punishment and extinction. One could cull similar quotations from any of Skinner's other books.

Beyond Freedom and Dignity was first publsihed in 1971. Did Skinner merely elaborate the findings of Miller, Olds, Breland and Breland and Sevenster? Did he adapt his theories to the fact that the concept of "reinforcement of behaviour by rewards" does not apply to all actions or to all organisms? Does he take cognizance of his failure to offer a proper definition of "reward"? Does he mention the fact that even the distinction between "classical" and "operant" conditioning has been challenged? Nothing of the kind. In this book, the old Skinner gospel is simply repeated: there must be more positive reinforcement of behaviour and less negative reinforcement. People must be praised not punished. Those who do good should be reinforced (offered rewards); those who do evil should receive neither punishment nor reward.

It is obvious that Skinner would not have written as he did had he worked with racoons instead of rats. Had he done so, he would never have taught his daughter with the aid of a large Skinner box, would never have said that a piano is a Skinner box with eighty-eight keys.

Skinner sees the whole of human life through the bars of his box. Because he writes with great conviction and extremely well and also because there are thousands of people who hanker after simplistic and unequivocal explanations of existence, it is understandable that he should have a large following. Moreover, some of the learning machines devised by him have prodouced excellent results, especially with deaf, dyslexic and other children with educational handicaps. Such children respond particularly well if they are reinforced with rewards to which they do not become habituated too quickly. Praises are, by their very nature, not nearly as suitable as, say, presenting the child with a small block that he may use to complete a construction.

What human actions does the behaviour of the rat in the Skinner box resemble? I have thought about it for a long time and have been quite unable to find anything in human behaviour that works according to an all-or-nothing concept of reward or no reward, a process in which the reward is constant and yet not subject to habituation. What probably comes closest to this process is the solving of jigsaw puzzles. If the child adds a piece correctly, he is rewarded: the piece fits. The child cannot tell in advance if the piece will fit (the variable ratio). His chief motive is not to complete the puzzle, for as soon as he has done so there is a good chance that he will jumble the pieces up and start all over again. If he fails to find fitting pieces after a long search he will abandon the puzzle (extinction). If he is rewarded now and then, but not too often, with a fitting piece, he will persist fanatically, forgetting food and drink, i.e. he does not become habituated. Adult puzzlers often behave in the same way.

Few people would go so far as to look upon all human behaviour as the solving of jigsaw puzzles. What happens when they do is something you can read in Skinner's *Science and Human Behaviour* and in his *Beyond Freedom and Dignity*, and what it can lead to you may gather from his *Walden Two*, a novel about a society in which everyone is content because the government ensures their happiness by a system of adequate rewards. When that book first appeared, many people declared that they did not wish to live in *Walden Two*. Skinner could not understand why. In fact, however, his *Walden Two* is nothing but Aldous Huxley's *Brave New World* or the happy state George Orwell has described so impressively in his *1984*.

IN CONCLUSION

In India the rat is the mount of Ganesha, the beloved god with the elephant's head. In our culture, the rat will never replace the ass as so exalted an animal. However, there is nothing to stop us from revising our general opinion of it. As long, moreover, as we continue our shameless exploitation and destruction of nature, we have no right to point a finger at other creatures. Rats, too, are living beings, precious animals. A wild rat may not be friendly, but its descendants are easy to handle and their progeny will eat from one's hand and can be kept as domestic pets. In this book I have tried to convey a clearer picture of the animal we call the rat. All those facets that did not contribute directly to this picture have been drawn very sketchily. Thus I have said nothing about the rise and fall of the guild of ratsbane vendors and rat-catchers whose interesting history would justify a separate book. In so short a work, I have also been unable to dwell on rats in legends, literature and paintings at any length.

The photographs are intended to provide as complete as possible a catalogue of the behaviour of brown rats. There are, however, some forms of behaviour that involve so many quick movements that they cannot, by virtue of this fact, be captured on still photographs.

I lack the space to acknowledge all the books, papers and journals that I have consulted, but most of the works listed below contain extensive bibliographies, to which the reader is referred. Nor can I acknowledge all the help I have received in writing this book; suffice it to express my special thanks to Dr A. M. Husson, my guide to the complicated classification of rodents, to A. J. Ophof, who has kindly allowed me to publish a number of photographs from *Rat en Muis*, and to Professor Dr P. Sevenster and my wife who read the manuscript and made many valuable suggestions for improvements in its form and content.

BIBLIOGRAPHY

S. A. Barnett, *The Rat*, London, Methuen & Co. Ltd. 1963.

G. E. H. Barrett-Hamilton and M. A. C. Hinton, *A History of British Mammals*, Parts 18 and 19, London, Gurney & Jackson, 1916.

K. Becker and H. Kemper, *Der Rattenkönig*, Berlin, Duncker & Humblot, 1964.

D. Chitty, *Control of Rats and Mice*, Parts 1 and 2, Oxford, Clarendon Press, 1954.

J. Cornelissen, *De muizen en ratten in de folklore*, Antwerp, Gust Janssens, 1923.

J. Ellerman, R. W. Hayman and G. W. C. Holt, *The Families and Genera of Living Rodents*, London, British Museum, 1941.

R. F. Ewer, *The Biology and Behaviour of a Free-living Population of Black Rats*, Animal Behaviour Monographs, London, 1972.

E. C. Grant, "An analysis of the social behaviour of the male laboratory rat", in *Behaviour* 21: 260–81, Leyden, 1963.

R. Hinde, *Animal Behaviour*, London, McGraw-Hill Book Company, 1970.

N. L. Munn, *Handbook of Psychological Research on the Rat*, New York, Houghton Mifflin, 1950.

B. F. Skinner, *The Behavior of Organisms*, New York, Appleton Century, 1938.

F. Steiniger, *Rattenbiologie und Rattenbekämpfung*, Stuttgart, 1952.

H. J. Telle, "Beitrag zur Kenntnis der Verhaltensweise von Ratten vergleichend dargestellt bei *Rattus norvegicus* und *Rattus rattus*", *Zeitschrift für Angew, Zoologie*, 53, 1966, pp. 129–96.

G. Ungar, *Molecular Mechanisms in Memory and Learning*, New York-London, Plenum Press, 1970.

H. Zinsser, *Rats, Lice and History*, 1935.

Index of Names

This book must be returned by the date specified at the time of issue as the DATE DUE FOR RETURN.
The loan may be extended (personally, by post, telephone or online) for a further period if the book is not required by another reader, by quoting the above number / author / title.

Enquiries: 01709 336774

www.rotherham.gov.uk/libraries